ELECTRICIAN'S RADIO ANATOMY MANUAL

GOAT VS BAPHOMET:
THEY NOT LIKE US

Chase Duquesnay
Dr. EnQi ReaL

Amazon

Copyright © 2024 Chase Duquesnay

All rights reserved

The characters and events portrayed in this book are fictitious. Any similarity to real persons, living or dead, is coincidental and not intended by the author. No part of any of the books written by these authors or in these series is meant to be taken as medical advice! These are strictly edutainment.

No part of this book may be reproduced, or stored in a retrieval system, or transmitted in any form or by any means, electronic, mechanical, photocopying, recording, or otherwise, without express written permission of the publisher.

ISBN: 9798300973414

Cover design by: Enqi Real

Printed in the United States of America

to those who love Osiris & Jesus!

CONTENTS

Title Page
Copyright
Dedication
Introduction

GOAT	1
BOOK OF HORNS	58
Stars	142
Iron	166
OCCUPATION	220
DR.EnQi	235
Technology Equipment Recap	250
Melanin, Radiation, and Energy Transduction in Fungi	271
Analysis of Melanin Properties in Radio-Frequency Range Based on Distribution of Relaxation Times	287

INTRODUCTION

All I can say is dive in, there is too much to cover in a quick intro. I can say that God wants you closer to him, so there is a lot to take in.

The plants are alive, yes and discussing your every move. The Air is alive, recording your every word and deed. That is Nature and God, Satan has retaliated with Technology. He is also watching and recording your every word and deed.

You have to pick a side, you can not fake it anymore. The reason so many christians want to be gay and christian is simple. They aren't really Christians. You know how many muslims eat pork everyday? They are muslim because they family was muslim or they went to prison etc... Their hearts don't belong to Christ, Osiris, Buddha, Allah, John, Mary, Isis none of that! They just use other 'religious' ideologies' as a tool

to dispute whatever is in front of them. Their hearts belong to Satan, they secretly are apart of the Do as Thou Whilst cult!
Beverly
Blavatsky
Crowley
Pasons
LaVey
Oh my...

When people are mislead into believing they have a unlimited sin debit card, they cool being in church, masjid, temple, synagogue etc... Once they gotta eat right, walk right, sex right etc... They like God aint real anyway lol....

I have the same beliefs with Health as I do with Religion, informed decisions. You can do what you want but....

Horny - late 14c., "**made of horn**," from horn (n.) + -y (2). From 1690s as "callous, **resembling horn**." The colloquial meaning "lustful, sexually aroused," was in use certainly by 1889, perhaps as early as 1863; it probably derives from the late 18c. slang expression to have the horn, suggestive of male ~~sexual excitement (but eventually applied to women as well)~~; see horn (n.). As a noun it once also was a popular name for a **domestic cow**. For an adjective in the original sense of the word, hornish (1630s) and horn-like (1570s) are available.

Do you know the Horns of the Altar are mentioned **26** times in the Bible? Do you understand the significance of numbers in the Bible? Especially that number, **26**?

That number should have it's own cult following (pun intended). Its funny because made of Horn and the Domesticated Cow will play roles in this book.

My goals in this book, though they maybe hard to see sometime are to:
1) bring people to God & the Bible
2) get rid of the negative or uncool stigma associated with the bible
3) drop the veil between the 'Israelites' and the Kemetic people.
4) highlight the fact the Devil is just as real as God is, and highlight the hard work Satanists are doing everyday
5) teach the hidden science in the Bible & Kemetic text
6) pull up the skirt of modern atheist as the use religious intellectual property as their own I.P....
7) rub everyone's nose in the fact that I found the cause and cure of Heart Disease in the Bible

I have to say that although I discovered the cause and cure of Heart Disease in the Bible, I have definitely been studying the Heart in Kemetic text much longer. If not for my constant study of the weighing of the Heart Ceremony and the Ebers papyrus I wouldve never been able to do it.

We have a whole book called the Clean Abs Fascia book, based totally around the anatomy of Shu's Feather, you may know of it as Maat's Feather or through Hebrew as the Yod. It is one of my favorite discoveries, nooooo.... I did discover the Feather on the walls of Kemet or as the Yod, those props go elsewhere. I discovered them through study as function aspects of biochemistry, in fact I discovered that Kemetic text, the Bible and Masonry are all teaching Biochemistry...

In the book your holding we talk a good bit about Shu... Therefore I will show you what I discovered about the Fascia and Shu. I basically discovered that the Scale was a metaphor for balancing Magnetism against Electricity. The heart representing Magnetism and the Feather representing Electricity. From Jujitsu I recognized the Body as a system of lever scales, so I knew the feather had to have more than just a metaphorical meaning... Boooommm you gotta go read the book for more, it's in the Clean Abs Fascia book. Ill show you though so you don't think I am crazier than I already am. Shu's feathers make up what we call collagen.

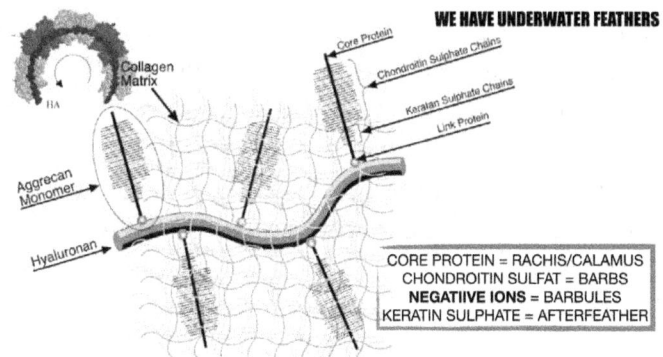

Schematic Presentation of Cartilage Extracellular Matrix.

I couldnt believe it, the most over looked aspect in health is the Fascia, in nutrition is collagen and there was the buried treasure. The Fascia is the reason we are able to convert calisthenics into a 96% reduction in cardiovascular disease risk!

My biggest pet pieve with our people when it comes to the Bible is dead study. 99% of the people that claim to study the Bible are liars, they just watch videos of people that study. Of the 1% of people that are actively studying the Bible, its dead study. Dead Study meaning they aren't trying to use any of the prayers, advice, dietary regiments, self improvements to see how this stuff impacts their lives. They are just studying to argue??? Then on the flipside you have good Christians, Muslims, Jews, Israelites, Buddhists etc... that don't study their material, they just try their best to live out the principles. That's cool but they aren't wearing their armor. You have to know that the two groups of people that are persecuted the most are Christians and 'Blacks'... If P then Q! If you are a Black Christian, you need to study, know at least the parts of the Bible associated with how you live. Here is an example:

1 Corinthians 9:24-25

24 Know ye not that they which run in a race run all, but one receiveth the prize? So run, that ye may obtain.

25 And every man that striveth for the mastery is temperate in all things. Now they do it to obtain a corruptible crown; but we an incorruptible.

1 Corinthians 3:17

17 If any man defile the temple of God, him shall God destroy; for the temple of God is holy, which temple ye are.

Daniel 1:10-16

10 And the prince of the eunuchs said unto Daniel, I fear my lord the king, who hath appointed your meat and your drink: for why should he see your faces worse liking than the

children which are of your sort? then shall ye make me endanger my head to the king.

11 Then said Daniel to Melzar, whom the prince of the eunuchs had set over Daniel, Hananiah, Mishael, and Azariah,

12 Prove thy servants, I beseech thee, ten days; and let them give us pulse to eat, and water to drink.

13 Then let our countenances be looked upon before thee, and the countenance of the children that eat of the portion of the king's meat: and as thou seest, deal with thy servants.

14 So he consented to them in this matter, and proved them ten days.

15 And at the end of ten days their countenances appeared fairer and fatter in flesh than all the children which did eat the portion of the king's meat.

16 Thus Melzar took away the portion of their meat, and the wine that they should drink; and gave them <u>pulse</u>.

Now for these verses I can get into my bag and say, if you improve your VO2Max by just 1 point, you can extend your lifespan 20-40 years! This is clinically documented...

I have already written a entire book dedicated to pulse, right? The Orthorexia Pulse of the Culture book. It's right here in Daniel that made me begin to research the pulse family. What I found has changed thousands of lives the last couple decades!!! The entire concept of the 40 Day Fruit Fast comes from Daniel & YaShua.

Matthew 4

Then was Jesus led up of the Spirit into the wilderness to be tempted of the devil.

2 And when **he had fasted forty days and forty nights**, he was afterward an hungred.

3 And when the tempter came to him, he said, If thou be the Son of God, command that these stones be made bread.

4 But he answered and said, It is written, Man shall not live by bread alone, but by every word that proceedeth out of the mouth of God.

You didnt know all these years I was bringing you back into the fold of the Lord huh....LMAO!!!

1 Timothy 4:8

For bodily exercise profiteth little: but godliness is profitable unto all things, having promise of the life that now is, and of that which is to come.

When I first read that verse, I violated Timothy in my mind lol... With maturity and further study I learned that over exercising was just as dangerous as not working out. I learned about sleep. Sleep, deep sleep is how your body washes your neurons. When you get into deep sleep, the Glials cells shrinks so the CSF can flow and remove crap like Amyloid Beta proteins etc... I learned that Weed, Liquor, Pills, even Caffeine inhibited this process taking years off your life. If you wake up cranky or with a headache it's probably because you didn't hit that deep sleep.

Please do me a favor, we didn't just pop this stuff on you...
The Radio book...??? Did you overlook that book? L'Goat book, maybe you bought it but havent read it? We definitely get into Radio Waves in those books! Nothing can prepare you for what your about to read but...

This is a cold hard ice bucket challange type read, if you have been using words like Osiris, Egyptian & Pharaoh without really thinking to yourself... These terms don't exist in the Mdu

Ntr... You are not prepared for your abou to read, this a 5,000 year journey.

GOAT

We have to tackle this Goat thing immediately. The Goat is not about Devil worship, the Goat is not the symbol of a Non-Binary God of the Emerald Tablets. The Goat is a symbol of the Creator God in Ancient Egypt and in the modern sense, God of the Bible. Satanist take the signs and symbols of God and make them into evil things, even worse the tools of Power they scare people away from.

The Rainbow is now and always has been the sign of God's covenant with man. The sign is the substance, the sign is water and visible light, the substance is electromagnetism. At this point you don't see the connection I get that but you gone learn today.

Proverbs 27:27 ESV
There will be enough goats' milk for your food, for the food of your household and maintenance for your girls.

Daniel 8:1-27 ESV
In the third year of the reign of King Belshazzar

a vision appeared to me, Daniel, after that which appeared to me at the first. And I saw in the vision; and when I saw, I was in Susa the citadel, which is in the province of Elam. And I saw in the vision, and I was at the Ulai canal. I raised my eyes and saw, and behold, a ram standing on the bank of the canal. It had two horns, and both horns were high, but one was higher than the other, and the higher one came up last. I saw the ram charging westward and northward and southward. No beast could stand before him, and there was no one who could rescue from his power. He did as he pleased and became great. As I was considering, behold, a male goat came from the west across the face of the whole earth, without touching the ground. And the goat had a conspicuous horn between his eyes. ...

Matthew 25:32 ESV
Before him will be gathered all the nations, and he will separate people one from another as a shepherd separates the sheep from the goats.

Proverbs 27:26 ESV
The lambs will provide your clothing, and the goats the price of a field.

Ezra 6:17 ESV
They offered at the dedication of this house of God 100 bulls, 200 rams, 400 lambs, and as a sin offering for all Israel 12 male goats, according to the number of the tribes of Israel.

I know this is still not making any sense yet, I pray that through my mistakes and typos, my inexperience as a author the truth still reaches you. I need you, I need your support, God needs you!

Lets start with Khnum, Khnum doesn't get the respect that he deserves. Khnum is arguably the Goat (pun intended), I believe this is done on purpose. The early Kemetic Scholars were mostly liberal college students, these men and their students like Reggie or Jabari today, were placed among they people to keep them dumb. I hope the handful of words we write here, read here, do his legacy some justice. The story of Khnum you may never have heard before but... you know the story!

The hieroglyphic symbol hnm (☐) often appearing in Khnum's name is derived from the word hnmt, signifying "well", or "spring". His name can also be connected to a Semitic root meaning "sheep". Alternatively, the formation of the name can be interpreted as "the beloved divine being". Khnum is also often described with the term iw m hapy, meaning "the coming of the Nile". Additionally, he is called Khnum-Ra, representing his role in the Nile cataract as the soul of the sun-god, Ra. Khnum's positions and powers are described through various titles such as the "Creator god", "Potter god", "Lord of

Life", "Lord of the Field", "Lord of Esna", "the good protector", and "Lord of the crocodiles".

Over time, the Egyptian word khn.m was later created to mean "shape" or "build", akin to Khnum's divine powers in creation. His significance also led to early theophoric names of him, for children, such as Khnum-Khufwy "Khnum is my Protector", the full name of Khufu, builder of the Great Pyramid of Giza. Khnum is a recurring figure in numerous of the hymns within the temples at Elephantine and Esna, showcasing his significance. Distinctively, The Morning Hymn to Khnum aligns him with the gods Amun and Shu, venerating him as the "Lord of life" and attributing him the ability to shape the bodies of humans. He was worshipped during the 1st dynasty, and during the early centuries CE, he was depicted as either a man with a ram's head or a ram with horizontal horns. It is believed that he created humans using clay and a potter's wheel. Khnum's creative powers extended beyond the earth, as he was thought to mold the bodies and spirits of newborn children, granting them life and health. It's believed that he and the goddess **Heqet** helped with the birth of a child.

It is Khnum that is the original Hairy Potter! Khnum is often referred to as the "Potter God," where he is described as using clay from the banks of the Nile to mold human bodies. He is believed to create not only the physical forms of people but also their ka, or spiritual essence, which is essential for life and identity in Egyptian belief. His creative acts are not limited to humans; he is also credited with shaping the gods and various living beings, reflecting his significance in the broader creation narrative of the ancient Egyptians. In Egyptian mythology, various texts highlight Khnum's role as a creator deity. The Pyramid Texts from the Old Kingdom frequently mention him in the context of creation, asserting that he not only creates humans but also oversees their conception and birth. The relationship between Khnum and the Nile is crucial, as the annual inundation provides the clay he utilizes, further cementing his identity as a life-giving force connected to natural cycles.

If I am sounding repetitive, I mean too! Khnum is the first God written to have created man from the Soil. Khnum created man from the mud of the Nile River, imbued them with Shu. Why don't these goofy Kemetic Scholars tell you this? Their schools trained them out of it! When they debate wether Kemet is the land of the Black People or the Black sut, they never tell you it's a dumb conversation because the people were made of the soil. Black Fertile Soil produced Black Fertile People. We still have that story today in the Bible. We still have the mysterious Aura of the Earth called Shuman Resonance. The fact of the longevity of these mnemonic devices is not just what enlightened me. I feel reversing cancer, reversing blindness, diabetes, kidney failure, heart disease etc... were to teach me, help me, heal me and at the same time, share my story, to create resonance! A personal revelation of mine, God has given me many, but this one many people have struggled with. The Goat lives in you, not only did this Ancient story survive to reach us today, the story is based on truth. The ventricle system in your brain today, is the Goat. <u>Your mind is literally everything, the ventricle system not the Fat is the most important part of the brain. When Ancient Myth tells you a story, then modern religion repeats that story, then modern science echos that story... You know the saying where there is smoke, there is fire? This is lava, these are focused laser beams, this is beyond smoke! In fact let me show you something... Smoke is now a slang term and Naggars don't even know what they are saying.</u>

Smoke - late Old English smoca, smocca (rare) "visible fumes and volatile material given off by burning or smoldering substances," related to smeocan "give off smoke," from Proto-Germanic *smuk- (source also of Middle Dutch smooc, Dutch smook, Middle High German smouch, German Schmauch), from PIE root *smeug- "to smoke; smoke" (source also of Armenian mux "smoke," Greek smykhein "to burn with smoldering flame," Old Irish much, Welsh mwg "smoke").

The more usual noun was Old English smec, which became dialectal smeech. From late 14c. as "a puff, cloud, or column of smoke." Figurative use, of something unsubstantial," is by 1540s; in reference to an obscuring medium, 1560s.

There is no fyre without some smoke [Heywood, 1562]

In other forms the proverb dates to mid-15c.

Abusive meaning "**Black person**" is attested from 1913, American English. Smoke-eater "firefighter" is by c. 1930. Figurative phrase go up in smoke "be destroyed" (as if by fire) is from 1933 (an earlier figurative image was come to smoke, "come to nothing," c. 1600, with a different image in mind). Smoke-alarm "device giving warning of smoke" is by 1936; smoke-detector from 1957. The figurative smoke-filled room, in U.S. politics the imagined site of private deals that secure party nominations, was popularized in the 1920 presidential election, in the days of open conventions and chain-smoking.

Middle English smoken, from Old English smocian, in late Old English smokian, "produce smoke, emit smoke," especially as a result of burning, intransitive, from smoke (n.1). Compare Dutch, Middle Low German smoken; for German rauchen, see reek (v.).

The transitive meaning "drive out or away or into the open by means of smoke" is attested from 1590s. Of chimneys, etc., "admit smoke outward instead of drawing it upward," 1660s. The meaning "to apply smoke to, to cure (bacon, fish, etc.) by exposure to smoke" is attested from 1590s. In old slang, "to sneer at, mock" (c. 1700).

In reference to tobacco, "draw fumes from burning into the mouth and puff them out again," it is first recorded 1604 in James I's "Counterblast to Tobacco." Related: Smoked; smoking.

"cigarette," slang, 1882, from smoke (n.1). Also "opium" (1884). The meaning "a spell of smoking tobacco" is recorded from 1835.

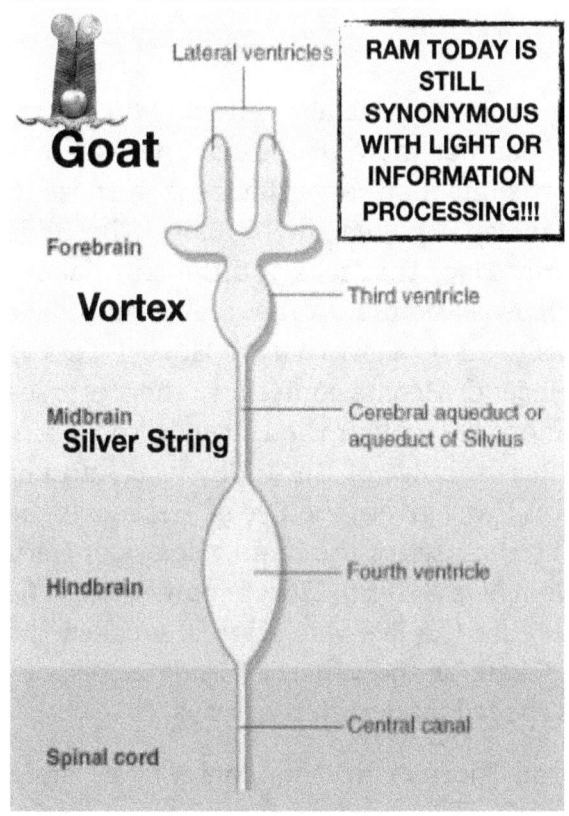

Your mind is literally everything, the ventricle system not the Fat is the most important part of the brain. When Ancient Myth tells you a story, then modern religion repeats that story, then modern science echos that story... You know the saying where there is smoke, there is fire? This is lava, these are focused laser beams, this is beyond smoke! The image of the Goat is built in to you, not only is the image of the Goat like a fingerprint, stamped into your anatomy, science today still refers to Memory and Information processing as RAM.

Not only from a particular veiw do you clearly see the Goat, but their are Horns! There are not only Horns but they serve the same purpose in your Brain as depicted in the Bible. The Bible is the Only Place that tells you the truth about Divine Horns. Divine Horns are Radio Wave antennas. This is Horn part... Horny! This is where Heqet or Heqate comes into the story, see the moment we look at the Horns in the brain, you will see the Eye of Heru. This was another revelation I was given, the Eye of Heru isn't the physical structures of the Brain, the eye of Heru is the Ventricle System. Hekate comes into the story because the Eye was originally called the Eye of Hekate or the Wheel of Heqate. Yes, her name is spelled many different ways.

The term "horns" in the context of brain

anatomy typically refers to the extensions of the lateral ventricles known as the frontal horn, temporal horn, and occipital horn. These structures are part of the ventricular system, which is crucial for the production and circulation of cerebrospinal fluid (CSF) within the brain. Understanding the anatomy and function of these "horns" helps illuminate their significance in both normal physiology and various neurological conditions.

Lateral Ventricles Overview

The lateral ventricles are a pair of C-shaped cavities located within the cerebral hemispheres. They consist of three primary regions: the body, the atrium, and the three horns (frontal, temporal, and occipital) that extend into corresponding lobes of the brain. The lateral ventricles communicate with the third ventricle via the interventricular foramina of Monro.

Frontal Horn

The frontal horn of the lateral ventricle is the most anterior extension. It begins at the foramina of Monro and extends forwards into the frontal lobe, curving along the frontal cortex. This region is bordered by the corpus callosum and contains important structures such as the anterior wall, formed by the genu of the corpus callosum, while the floor is defined by the rostrum. The frontal horn plays a role in a variety of cognitive functions, as it is situated in a critical area for language, problem-solving, and emotional regulation.

Temporal Horn

The temporal horn is the largest and longest of the three horns, extending laterally from the atrium into the temporal lobe. It is positioned below the thalamus and extends to the region of the amygdala. The floor of the temporal horn is formed by the collateral eminence and the hippocampus, while the roof is created by parts of the thalamus and the caudate nucleus. The temporal horn is associated with memory functions and emotional responses due to the proximity of critical structures like the hippocampus.

Occipital Horn

The occipital horn curves posteriorly and laterally from the atrium, extending into the occipital lobe. Its size can vary significantly among individuals and is important for visual processing as it is closely linked with the visual cortex. The occipital horn is less frequently discussed than the frontal and temporal horns, yet its structural integrity is vital for proper visual perception.

Clinical Significance

Anatomical variations in the horns of the

lateral ventricle can have clinical implications. For instance, asymmetries in size and shape can indicate underlying conditions, such as lateral ventricle enlargement due to neurodegenerative diseases or developmental anomalies. Radiologists pay close attention to the morphology of these structures during neuroimaging to differentiate between normal anatomical variants and pathological conditions.

Conclusion

The "horns" in the brain—frontal, temporal, and occipital—are critical components of the lateral ventricles, contributing to the overall function of the ventricular system. These structures not only play essential roles in the circulation of cerebrospinal fluid but also impact cognitive and sensory functions associated with the frontal, temporal, and occipital lobes. Understanding their anatomy and significance is essential for interpreting neurological health and diagnosing related conditions.

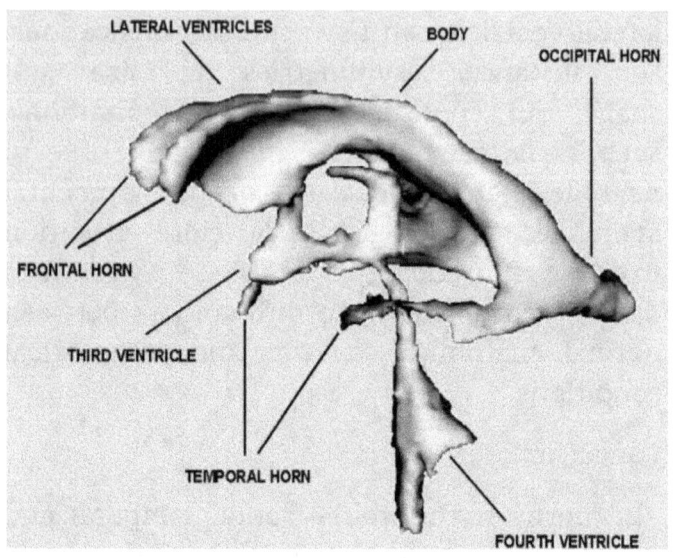

Anterior Horns, Posterior & the star of the show, the Temporal Horn. The inferior horn of the lateral ventricle, or temporal horn, is the largest of the horns. It extends anteriorly from the atrium beneath the thalamus and terminates at the amygdala. Time and Space are interpreted in this pathway of the Brain. Light, Sound, Time and Space only exist in your mind. The people that truly understand this, are using this knowledge to get rich and sacrifice the rest of the world.

Enter in the Clones. The Soulless people that copy and steal. Savages steal, mortals copy but only Gods Create! Eliphas Levi creates the Baphomet. The entire concept was to ge

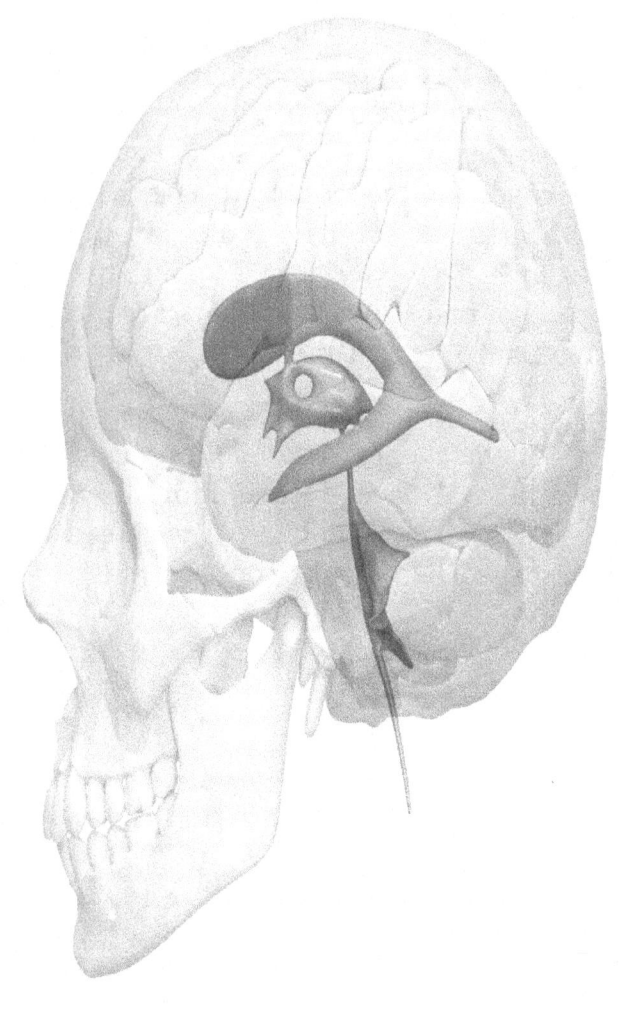

t   to   the   roots   of   man

ELECTRICIAN'S RADIO ANATOMY MANUAL

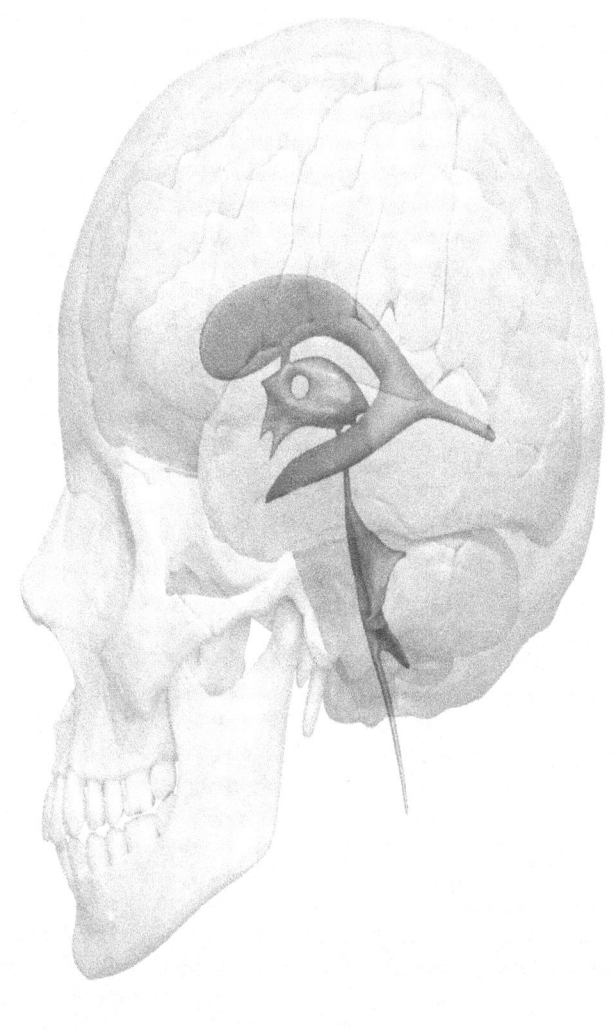

CHASE DUQUESNAY

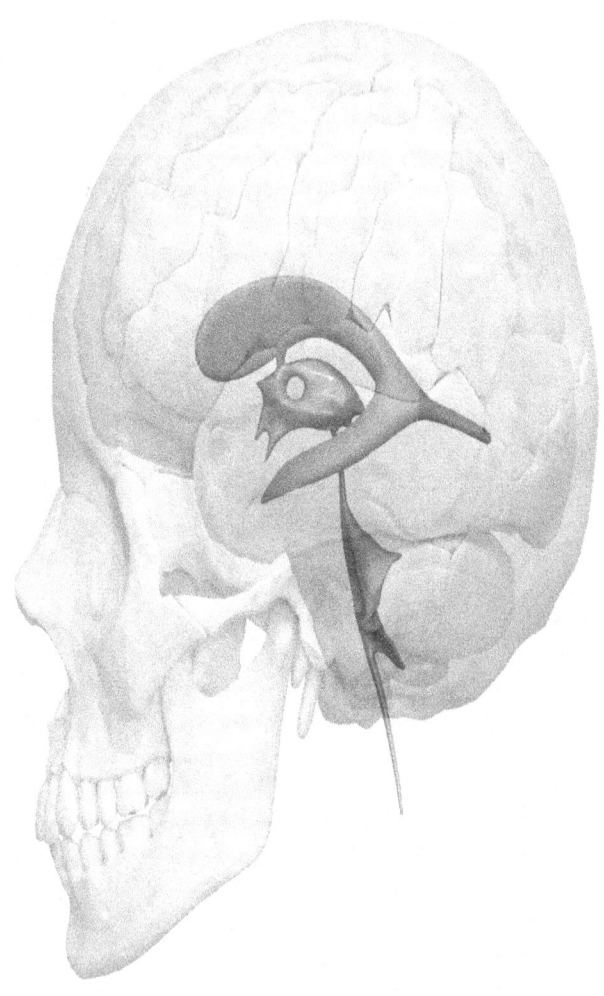

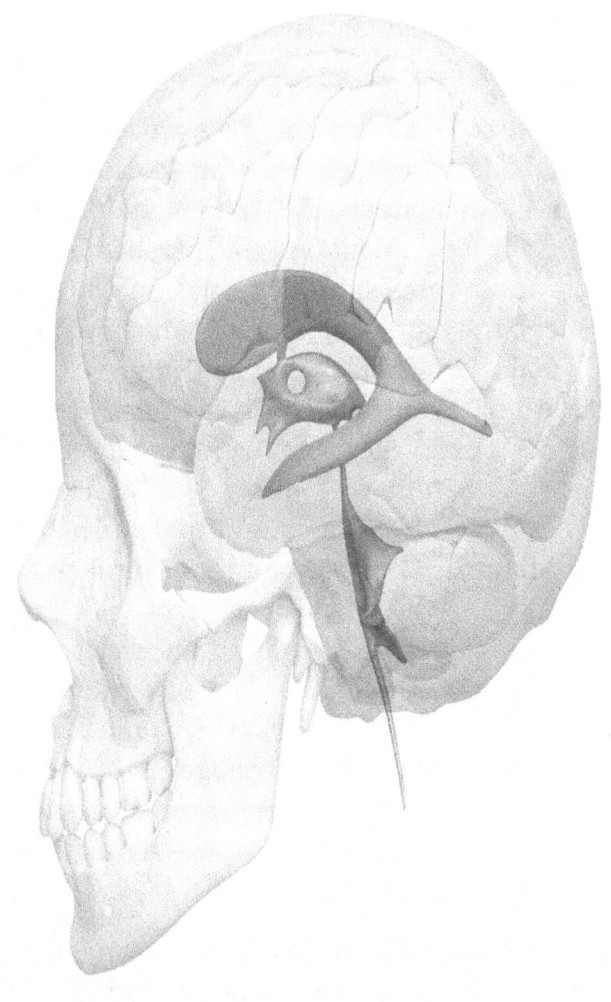

's psyche, it is this group of Europeans that turned the world against the Bible. Tt is this group of Europeans that turned the world against the Bible, and gave you drugs and sex instead. Understanding the brain is wired for God, that the Goat distributes opioids...

They took Freemasonry (Kemetic Science), Islam, straight Kemetic Science & the Upanishads and Devil worship, remixed them into a clean system of Satanism hidden from public view. The modes of membership and recognition were similar to Masonry (Kemetic Science), the name Baphomet is Muslim and the hid male on male anal sex under Chakra Activation. A chronicler of the First Crusade, reports that the troubadours used the term Bafomet for Muhammad, and Bafumaria for a mosque. Éliphas Lévi Zahed, born Alphonse Louis Constant (8 February 1810 – 31 May 1875), was a French esotericist, poet, and writer. Initially pursuing an ecclesiastical career in the Catholic Church, he abandoned the priesthood in his mid-twenties and became a ceremonial magician. At the age of 40, he began professing knowledge of the occult.[1] He wrote over 20 books on magic, Kabbalah, alchemical studies, and occultism.

The pen name "Éliphas Lévi", was an anagram of his given names "Alphonse Louis" into Hebrew. Of course your not officially a Satanist if you aren't practicing usery of God. That is the simplest way I can explain Satanism usery of God or God's Power. Eliphas or Alphonso would be considered a OG of Blavatsky's. Helena

Blavatsky took what Levi created to the next level, he can be considered the grandpa of Aleister Crowley! Levi -> Blavatsky -> Crowley -> Parsons -> You! Yep, you have been converted into a Devil worshipper. The Baphomet is not only a Scarecrow but to is the map of the soul and plan of attack for Satanist! Baphomet is the Androgynous God of the Emerald Tablets. Billy Carson, Nature Boy, Polight and many many others have helped usher in this cult secretly to the masses. These guys teach devil worship openly, they all began their 'conscious' missionary work as criminals. Jail was where they were co-opted. Billy Carson was allegedly a identity thief and bank fraud scammer, Polight was a petty thief that talked his friend into killing one of Dr. Yorks children so he could replace them, Natureboy was a molested Foster kid that began to sell his body in male strip clubs to men. These Satanist turned the culture of gangsters trying to become good men, to a culture of good men contracted into criminal indictments, they understand resonance. They contracted you into being a criminal and taught you how to make the kids feel you! They turned being Horny into sexual deviance, then made a sexual disease in the lab to kill you!

āutcharu (ātcharu) ⸻, part, or parts, of a chariot.

āutchatá (ātchatá) ⸻, Alt. K. 306 . . . .

āb ⸻, to be renowned, famous, strength (?)

āb ⸻, U. 270, ⸻, N. 719, horn, tusk of an elephant;

nuḥerḥer ⟨hieroglyphs⟩, N. 16, ⟨hieroglyphs⟩, P. 74, ⟨hieroglyphs⟩, M. 105, to rejoice.

nuḥes ⟨hieroglyphs⟩, negro; see ⟨hieroglyphs⟩

nukh ⟨hieroglyphs⟩, L.D. III, 140B, ⟨hieroglyphs⟩, to cook, to bake, to roast.

nus ⟨hieroglyphs⟩,

Ā [116]

plur. ⟨hieroglyphs⟩, U. 270, ⟨hieroglyphs⟩, N. 719; dual, ⟨hieroglyphs⟩, ⟨hieroglyphs⟩, Rougé, I.H. II, 114; ⟨hieroglyphs⟩ = Dhu'l Ḳarnên; ⟨hieroglyphs⟩, he with horns ready to gore; ⟨hieroglyphs⟩, U. 577, the four horns of the bull of Rā, the four horns of the world.

ābāti (?) ⟨hieroglyphs⟩, Thes. 1198, the gorer.

āb ⟨hieroglyphs⟩, tusk of ivory; see ab ⟨hieroglyphs⟩.

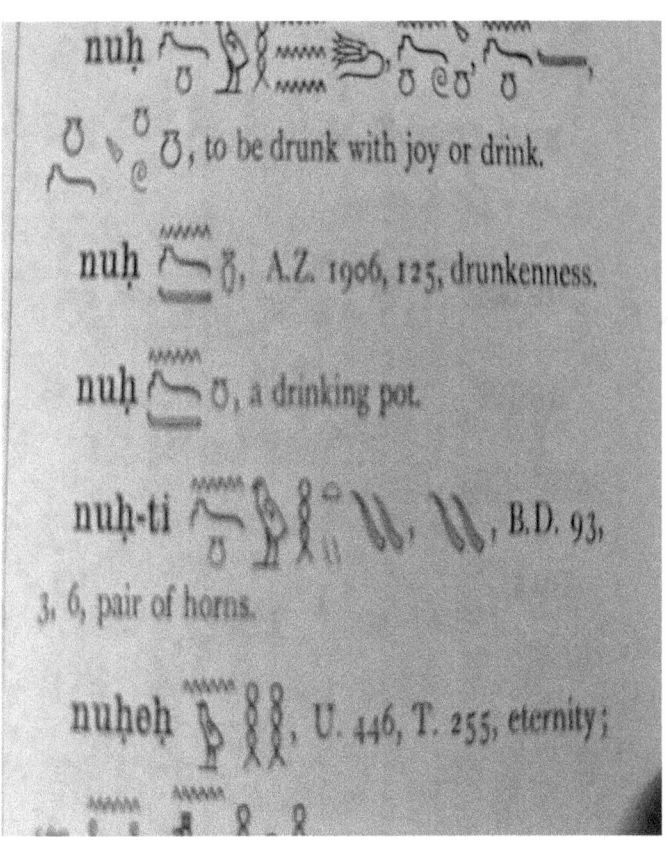

nuḥ ...., to be drunk with joy or drink.

nuḥ ...., A.Z. 1906, 125, drunkenness.

nuḥ ...., a drinking pot.

nuḥ-ti ...., B.D. 93, 3, 6, pair of horns.

nuḥeh ...., U. 446, T. 255, eternity;

mer ābu (?) [hieroglyphs], Anastasi IV, 3, 1, Koller Pap. 3, 1, inspector of horned cattle (?)

mer ābu ...... shu [hieroglyphs], inspector of horn, hoof, and feather, *i.e.*, overseer of all the cattle and feathered fowl; [hieroglyphs], Rec. 17, 4, inspector of horn, hoof, feather, and metal.

mer ānṭ [hieroglyphs], overseer of the storehouse.

mer u (?) [hieroglyphs], IV, 1115, [hieroglyphs]

The Horns of the Altar (mentioned **26** times)

Alert, Alter, AllTar, Altar, is, was, will be, the Mind!!! Whats the odds that the Divine Altar in the Bible has 4 horns, just like your Brain. Was the brain designed like the Altar or was the Altar designed like your Brain or is the Altar actually your Brain??? Let's first notice that this is the 37 chapter!

Exodus 37

1 Then Bezalel made the ark of acacia wood; two and a half cubits was its length, a cubit and a half its width, and a cubit and a half its height. 2 He overlaid it with pure gold inside and outside, and made a molding of gold all around it. 3 And he cast for it four rings of gold to be set in its four corners: two rings on one side, and two rings on the other side of it. 4 He made poles of acacia wood, and overlaid them with gold. 5 And he put the poles into the rings at the sides of the ark, to bear the ark. 6 He also made the mercy seat of pure gold; two and a half cubits was its length and a cubit and a half its width. 7 He made two cherubim of beaten gold; he made them of one piece at the two ends of the mercy seat: 8 one cherub at one end on this side, and the other cherub at the other end on that side. He made the cherubim at the two ends of one piece with the mercy seat. 9 The cherubim spread out their wings above, and covered the mercy seat with their wings. They faced one another; the faces of the cherubim were toward the mercy seat.

Making the Table for the Showbread

10 He made the table of acacia wood; two cubits was its length, a cubit its width, and a cubit and a half its height. 11 And he overlaid it with pure gold, and made a molding of gold all around it. 12 Also he made a frame of a handbreadth all around it, and made a molding of gold for the frame all around it. 13 And he cast for it four rings of gold, and put the rings on the four corners that were at its four legs. 14 The rings were close to the frame, as holders for the poles to bear the table. 15 And he made the poles of acacia wood to bear the table, and overlaid them with gold. 16 He made of pure gold the utensils which were on the table: its dishes, its cups, its bowls, and its pitchers for pouring.

Making the Gold Lampstand

17 He also made the lampstand of pure gold; of hammered work he made the lampstand. Its shaft, its branches, its bowls, its ornamental knobs, and its flowers were of the same piece. 18 And six branches came out of its sides: three branches of the lampstand out of one side, and three branches of the lampstand out of the other side. 19 There were three bowls made like almond blossoms on one branch, with an ornamental knob and a flower, and three bowls made like almond blossoms on the other branch, with an ornamental knob

ELECTRICIAN'S RADIO ANATOMY MANUAL

and a flower—and so for the six branches coming out of the lampstand. 20 And on the lampstand itself were four bowls made like almond blossoms, each with its ornamental knob and flower. 21 There was a knob under the first two branches of the same, a knob under the second two branches of the same, and a knob under the third two branches of the same, according to the six branches extending from it. 22 Their knobs and their branches were of one piece; all of it was one hammered piece of pure gold. 23 And he made its seven lamps, its wick-trimmers, and its trays of pure gold. 24 Of a talent of pure gold he made it, with all its utensils.

Making the Altar of Incense

25 He made the incense altar of acacia wood. Its length was a cubit and its width a cubit—it was square—and two cubits was its height. **Its horns were of one piece with it**. 26 And he overlaid it with pure gold: its top, its sides all around, and **its horns**. He also made for it a molding of gold all around it. 27 He made two rings of gold for it under its molding, by its two corners on both sides, as holders for the poles with which to bear it. 28 And he made the poles of acacia wood, and overlaid them with gold.

Making the Anointing Oil and the Incense

29 He also made the holy anointing oil and the

pure incense of sweet spices, according to the work of the perfumer.

Psalms 118

1 Oh give thanks unto the Lord; for he is good: because his mercy endureth for ever.

2 Let Israel now say, that his mercy endureth for ever.

3 Let the house of Aaron now say, that his mercy endureth for ever.

4 Let them now that fear the Lord say, that his mercy endureth for ever.

5 I called upon the Lord in distress: the Lord answered me, and set me in a large place.

6 The Lord is on my side; I will not fear: what can man do unto me?

7 The Lord taketh my part with them that help me: therefore shall I see my desire upon them that hate me.

8 It is better to trust in the Lord than to put confidence in man.

9 It is better to trust in the Lord than to put confidence in princes.

10 All nations compassed me about: but in the name of the Lord will I destroy them.

11 They compassed me about; yea, they compassed me about: but in the name of the Lord I will destroy them.

12 They compassed me about like bees: they are quenched as the fire of thorns: for in the name of the Lord I will destroy them.

13 Thou hast thrust sore at me that I might fall: but the Lord helped me.

14 The Lord is my strength and song, and is become my salvation.

15 The voice of rejoicing and salvation is in the tabernacles of the righteous: the right hand of the Lord doeth valiantly.

16 The right hand of the Lord is exalted: the right hand of the Lord doeth valiantly.

17 I shall not die, but live, and declare the works of the Lord.

18 The Lord hath chastened me sore: but he hath not given me over unto death.

19 Open to me the gates of righteousness: I will go into them, and I will praise the Lord:

20 This gate of the Lord, into which the

righteous shall enter.

21 I will praise thee: for thou hast heard me, and art become my salvation.

22 The stone which the builders refused is become the head stone of the corner.

23 This is the Lord's doing; it is marvellous in our eyes.

24 This is the day which the Lord hath made; we will rejoice and be glad in it.

25 Save now, I beseech thee, O Lord: O Lord, I beseech thee, send now prosperity.

26 Blessed be he that cometh in the name of the Lord: we have blessed you out of the house of the Lord.

27 God is the Lord, which hath shewed us light: bind the sacrifice with cords, **even unto the horns of the altar**.

28 Thou art my God, and I will praise thee: thou art my God, I will exalt thee.

29 O give thanks unto the Lord; for he is good: for his mercy endureth for ever.

Ezekiel 43

1 Afterward he brought me to the gate, even the

gate that looketh toward the east:

2 And, behold, the glory of the God of Israel came from the way of the east: and his voice was like a noise of many waters: and the earth shined with his glory.

3 And it was according to the appearance of the vision which I saw, even according to the vision that I saw when I came to destroy the city: and the visions were like the vision that I saw by the river Chebar; and I fell upon my face.

4 And the glory of the Lord came into the house by the way of the gate whose prospect is toward the east.

5 So the spirit took me up, and brought me into the inner court; and, behold, the glory of the Lord filled the house.

6 And I heard him speaking unto me out of the house; and the man stood by me.

7 And he said unto me, Son of man, the place of my throne, and the place of the soles of my feet, where I will dwell in the midst of the children of Israel for ever, and my holy name, shall the house of Israel no more defile, neither they, nor their kings, by their whoredom, nor by the carcases of their kings in their high places.

8 In their setting of their threshold by my

thresholds, and their post by my posts, and the wall between me and them, they have even defiled my holy name by their abominations that they have committed: wherefore I have consumed them in mine anger.

9 Now let them put away their whoredom, and the carcases of their kings, far from me, and I will dwell in the midst of them for ever.

10 Thou son of man, shew the house to the house of Israel, that they may be ashamed of their iniquities: and let them measure the pattern.

11 And if they be ashamed of all that they have done, shew them the form of the house, and the fashion thereof, and the goings out thereof, and the comings in thereof, and all the forms thereof, and all the ordinances thereof, and all the forms thereof, and all the laws thereof: and write it in their sight, that they may keep the whole form thereof, and all the ordinances thereof, and do them.

12 This is the law of the house; Upon the top of the mountain the whole limit thereof round about shall be most holy. Behold, this is the law of the house.

13 And these are the measures of the altar after the cubits: The cubit is a cubit and an hand breadth; even the bottom shall be a cubit, and

the breadth a cubit, and the border thereof by the edge thereof round about shall be a span: and this shall be the higher place of the altar.

14 And from the bottom upon the ground even to the lower settle shall be two cubits, and the breadth one cubit; and from the lesser settle even to the greater settle shall be four cubits, and the breadth one cubit.

15 **<u>So the altar shall be four cubits; and from the altar and upward shall be four horns</u>**.

16 And the altar shall be twelve cubits long, twelve broad, square in the four squares thereof.

17 And the settle shall be fourteen cubits long and fourteen broad in the four squares thereof; and the border about it shall be half a cubit; and the bottom thereof shall be a cubit about; and his stairs shall look toward the east.

18 And he said unto me, Son of man, thus saith the Lord God; These are the ordinances of the altar in the day when they shall make it, to offer burnt offerings thereon, and to sprinkle blood thereon.

19 And thou shalt give to the priests the Levites that be of the seed of Zadok, which approach unto me, to minister unto me, saith the Lord God, a young bullock for a sin offering.

20 And thou shalt take of the blood thereof, and put it on the four horns of it, and on the four corners of the settle, and upon the border round about: thus shalt thou cleanse and purge it.

21 Thou shalt take the bullock also of the sin offering, and he shall burn it in the appointed place of the house, without the sanctuary.

22 And on the second day thou shalt offer a kid of the goats without blemish for a sin offering; and they shall cleanse the altar, as they did cleanse it with the bullock.

23 When thou hast made an end of cleansing it, thou shalt offer a young bullock without blemish, and a ram out of the flock without blemish.

24 And thou shalt offer them before the Lord, and the priests shall cast salt upon them, and they shall offer them up for a burnt offering unto the Lord.

25 Seven days shalt thou prepare every day a goat for a sin offering: they shall also prepare a young bullock, and a ram out of the flock, without blemish.

26 Seven days shall they purge the altar and purify it; and they shall consecrate themselves.

27 And when these days are expired, it shall be,

that upon the eighth day, and so forward, the priests shall make your burnt offerings upon the altar, and your peace offerings; and I will accept you, saith the Lord God.

In L'Goat book, and the Electrician Prayer Manual, we discuss the construction of Great Pyramid (built by the son of Khnum), the House of Light. The Temple of Solomon is a model of the Great Pyramid based on the numbers 26 & 137.

Iron is the 26th element, God's number is 26 and horns of the altar is mentioned 26 times!!! Men lie, women lie, numbers don't!

THE GREAT PYRAMID WAS A RADIO STATION! Please read &/or reread the Movement book! The Great Pyramid is literally built where the 3rd Ventricle is in your brain. The Pyramid is a repository of information, the Kings Chamber has the 137 signature built in. Keep in mind Radio Waves give birth to what we know of as Electromagnetism. We know that Serotonin (speed) and Dopamine (slow) regulate our perception of time. We know space and time, represented by the Pyramid are created via the Entorhinal Cortex with the Hippocampus & Amygdala. Time, Space, Light and Sound are all in your mind, The largest impact on how that happens is Radio Waves!

We are sold the moon story, to justify and everything create a wall of illusion. The Space Agencies around the world are just creating reality. They are spending billions on shooting Radio waves at foreign Stars! The Radio Waves are aimed at us, aimed at our minds, aimed at our atmosphere!

A theoretical study has modelled how radio waves behave when passing through the Great Pyramid of Giza in Egypt. Mikhail Balezin and colleagues at St Petersburg's ITMO University in Russia and Germany's Laser Zentrum Hannover used multipole analysis to approximate how the electromagnetic waves would be influenced by the famous landmark. As well as offering a new way to study interiors of huge structures, the technique is also being used to characterize pyramidal nanoparticles.

The interior of the Great Pyramid has been probed using various forms of radiation including cosmic muons. Indeed, the muon study has found evidence for a previously unknown chamber buried deep within the iconic structure.

Now, Balezin and colleagues have performed the first study of how the pyramid would interact with radio waves. They constructed a numerical model to simulate the behaviour of radio waves with wavelengths of 200-600 m as they passed through a virtual pyramid. Such wavelengths were chosen because they are slightly longer than the physical dimensions of the Great Pyramid, which is about 140 m tall and measures 230 m along each of its four sides.

Solid limestone

The team first modelled the pyramid as solid limestone with no internal chambers. Then, they looked at how the presence of chambers would affect the radio waves. Their simulations predict that the chambers act as resonators, concentrating electromagnetic energy inside the chambers. They also found that the pyramid as a whole focussed radio waves incident from above into a region just below the structure.

The team worked-out that some incident waves would be scattered by internal structures and that others would be absorbed. They were also able to map the distribution of electromagnetic fields inside the pyramid.

The simulations used multipole analysis – a mathematical technique that can approximate interactions between complex objects and electromagnetic fields. The technique involves replacing the object with much simple set of radiation emitters known as multipoles. With a knowledge of the properties of each individual multipole, the researchers could use mathematical functions to approximate how their combined emitted field would be scattered.

Similar scales
Normally, the team studies interactions between light and nanoparticles – where the wavelength of the light is also larger than the size of the structures of interest. This similarity inspired Balezin and colleagues to look at pyramids and show that on very different length scales of nanometres and hundreds of metres the scattering of electromagnetic waves ultimately depends on the size, shape and refractive index of the objects.

The team is now looking at how pyramidal nanoparticles can be used in new and innovative ways to create new technologies such as nanosensors and highly efficient solar cells. The team also plans to do further simulations of the Great Pyramid using radio waves at shorter wavelengths.

Electromagnetic properties of the Great Pyramid: First multipole resonances and energy concentration

Mikhail Balezin; Kseniia V. Baryshnikova; Polina Kapitanova ; Andrey B. Evlyukhin

Resonant response of the Great Pyramid interacting with external electromagnetic waves of the radio frequency range (the wavelength range is 200–600 m) is theoretically investigated. With the help of numerical simulations and multipole decomposition, it is found that spectra of the extinction and scattering cross sections include resonant features associated with excitation of the Pyramid's electromagnetic dipole and quadrupole moments. Electromagnetic field distributions inside the Pyramid at the resonant conditions are demonstrated and discussed for two cases, when the Pyramid is located in a homogeneous space or on a substrate. It is revealed that the Pyramid's chambers can collect and concentrate electromagnetic energy for the both surrounding conditions. In the case of the Pyramid on the substrate, at the shorter wavelengths, the electromagnetic energy accumulates in the chambers providing local spectral maxima for electric and magnetic fields. It is shown that basically the Pyramid scatters the electromagnetic waves and focuses them into the substrate region. The spectral dependence of the focusing effect is discussed.

Auditory response to pulsed radiofrequency energy
J A Elder 1, C K Chou

Abstract
The human auditory response to pulses of radiofrequency (RF) energy, commonly called

RF hearing, is a well established phenomenon. RF induced sounds can be characterized as low intensity sounds because, in general, a quiet environment is required for the auditory response. The sound is similar to other common sounds such as a click, buzz, hiss, knock, or chirp. Effective radiofrequencies range from 2.4 to 10000 MHz, but an individual's ability to hear RF induced sounds is dependent upon high frequency acoustic hearing in the kHz range above about 5 kHz. The site of conversion of RF energy to acoustic energy is within or peripheral to the cochlea, and once the cochlea is stimulated, the detection of RF induced sounds in humans and RF induced auditory responses in animals is similar to acoustic sound detection. The fundamental frequency of RF induced sounds is independent of the frequency of the radiowaves but dependent upon head dimensions. The auditory response has been shown to be dependent upon the energy in a single pulse and not on average power density. The weight of evidence of the results of human, animal, and modeling studies supports the thermoelastic expansion theory as the explanation for the RF hearing phenomenon. RF induced sounds involve the perception via bone conduction of thermally generated sound transients, that is, audible sounds are produced by rapid thermal expansion resulting from a calculated temperature rise of only 5×10^{-6}

degrees C in tissue at the threshold level due to absorption of the energy in the RF pulse. The hearing of RF induced sounds at exposure levels many orders of magnitude greater than the hearing threshold is considered to be a biological effect without an accompanying health effect. This conclusion is supported by a comparison of pressure induced in the body by RF pulses to pressure associated with hazardous acoustic energy and clinical ultrasound procedures.

Life

Human ear inspires universal radio chip

10 June 2009
A COMPUTER chip modelled on the human ear could be used in universal receivers for radio-frequency signals ranging from cellphone and wireless internet transmissions to radio and television broadcasts.

Devices such as cellphones or FM radios are generally tuned to only a narrow frequency band. The new device is inspired by the network of hairs in the inner ear, which can pick up a wide range of sound frequencies.

We can hear because sound waves make the eardrum vibrate, which creates waves in...
Hearing radio frequencies
I was reading the Wikipedia article on tinnitus, and came across this pearl of a sentence:

A common and often misdiagnosed condition

that mimics tinnitus is Radio Frequency (RF) Hearing in which subjects have been tested and found to hear high-pitched transmission frequencies that sound similar to tinnitus.

Hmm, what? Yes, humans, under special circumstances, can hear radio-frequency pulses in the range of 2.4MHz to 10GHz (corresponding to radio frequencies and microwave) as buzzes, clocks, hiss or knocking at apparent auditory frequencies of 5kHz and higher (very high-pitched). That doesn't mean that you can hear talk radio by receiving AM waves; it just means that when it's very very quiet, you can hear a faint high-pitched noise from RF sources.

But how could electromagnetic waves be detected as sound, which is a pressure wave? After all, light is an EM wave too, but we don't hear light! It's a long story, but basically, you're a microwave bongo head. Elder and Chou (2003) offer a thorough overview of the phenomenon.

RF hearing was first reported in the 1940s by people working with radar, but reports were dismissed as illusions or hallucinations. The phenomenon was investigated scientifically by Frey in 1961, who concluded that RF hearing is a real thing. It can be stopped, for example, by placing a piece of aluminum between the RF source and the ear.

RF sources can only be heard by people with

working audition above 5kHz. This would imply that RF sources create an acoustic vibration close to the cochlea that gets detected as high-frequency sound. Indeed, one can record electrical potentials inside the cochlea evoked by RF pulses that look just like potentials evoked by sound waves.

The authors further report that the apparent acoustic frequency of the RF pulse is independent of the EM frequency of the actual pulse but dependent upon head dimensions. So EM energy gets absorbed by the head and somehow this energy is transformed into pressure waves that get reshaped by the head. Thus, microwave bongo head.

Doc, what does all that have to do with Horns in Egypt or in the Bible? Good Question.

HORN - Old English horn "horn of an animal; projection, pinnacle," also "wind instrument" (originally one made from animal horns), from Proto-Germanic *hurni- (source also of German Horn, Dutch horen, Old Frisian horn, Gothic haurn), from PIE root *ker- (1) "horn; head."

Late 14c. as "one of the tips of the crescent moon." The name was retained for a class of musical instruments that developed from the hunting horn; the French horn is the true representative of the class. Of dilemmas from

1540s; of automobile warning signals from 1901. Slang meaning "erect penis" is suggested by c. 1600. Jazz slang sense of "trumpet" is by 1921. Meaning "telephone" is by 1945. Figurative senses of Latin cornu included "salient point, chief argument; wing, flank; power, courage, strength." Horn of plenty is from 1580s. To make horns at "hold up the fist with the two exterior fingers extended" as a gesture of insult is from c.1600.

Symbolic of cuckoldry since mid-15c. (the victim was fancied to grow one on his head). The image is widespread in Europe and perhaps as old as ancient Greece. The German linguist Hermann Dunger ('Hörner Aufsetzen' und 'Hahnrei', "Germania" 29, 1884) ascribes it to a custom surviving into 19c., "the old practice of engrafting the spurs of a castrated cock on the root of the excised comb, which caused them to grow like horns" [James Hastings, "Encyclopedia of Religion and Ethics"] but the image could have grown as well from a general gesture of contempt or insult made to wronged husbands, "who have been the subject of popular jest in all ages" [Hastings].

1690s, "to furnish with horns," from horn (n.). Earlier in figurative sense of "to cuckold" (1540s). Meaning "to push with the horns" (of cattle, buffalo, etc.) is from 1851, American English; phrase horn in "intrude" is by 1880,

American English, originally cowboy slang. Related: Horned; horning.

ANTENNA - 1690s, "<u>to furnish with horns</u>," from horn (n.). Earlier in figurative sense of "to cuckold" (1540s). Meaning "to push with the horns" (of cattle, buffalo, etc.) is from 1851, American English; phrase horn in "intrude" is by 1880, American English, originally cowboy slang. Related: Horned; horning.

*Before we keep going do you see that Antenna and Horn were once the same thing???? In fact it looks like horn was first. Horn was the word for natural Antenna or thing that sends and receives Radio Waves. Just stop reading for a second, go back to every weird Hieroglyph and every weird chapter you have read in the Bible, they all instantly make sense once you swap out musical instrument for antenna type vibe.

THERE IS EVEN A THING CALLED A HORN ANTENNA WHICH HAS BEEN IN USE HUNDREDS OR THOUSANDS OF YEARS!!! He who controls the sky, controls the people! Space Agencies control space, space is formed in your mind! Controlling time and space means controlling your mind! I am praying I am saying the right things, like using the right words to get you to understand what I am saying.

The word, creates. Technology monitors,

supports and sustains, what the word has created. Let's explore the concept of being High, we use high most in two spaces: Drugs and God. The most High and get High. When we talk high, we have to ask ourselves is the high reality the real one or is the sober reality the real one. I can say plainly the sober mind, is the mind in agreeance. Once you are high, you are in your own headspace. Depending on the drug, you may even start hallucinating. In the high headspace those hallucinations are REAL. There maybe other entities here, entities that maybe older or mentally stronger than us that do not have Flesh or Flesh like ours.

They Not Like Us and I am not talking about white, yellow, brown... I am talking about the full court press to get us to accept Aliens!!!

Alien - c. 1300, "strange, **foreign**," from Old French alien "strange, foreign;" as a noun, "an alien, stranger, foreigner," from Latin alienus "of or belonging to another, **not one's own**, foreign, strange," also, as a noun, "a stranger, foreigner," adjective from alius (adv.) "another, other, different" (from PIE root *al- (1) "**beyond**").

The meaning "residing in a country not of one's birth" is from mid-15c. The sense of "**wholly different in nature**" is from 1670s. The meaning "not of this Earth" is recorded by 1920.

An alien priory (mid 15c.) is one owing obedience to a religious jurisdiction in a foreign country.

"foreigner, citizen of a foreign land," early 14c., from alien (adj.) or from noun use of the adjective in French and Latin. In the science fiction sense "being from another planet," from 1953.

I mean I don't believe in coincidence, HOV & Beyonce. Those names together take the concept of power couple to another level. Jay from Brooklyn, Im from the Bronx, but I am also Haitian, just like Beyonce! Never forget that, Beyonce is a Haitian. They are the dangerous Haitians too, not saying bad or good, the jury is out... I just saying, don't sleep and think J calling all the shots over there. Anyway back to business.

<u>**A horn antenna or microwave horn is an antenna that consists of a flaring metal waveguide shaped like a horn to direct radio waves in a beam**</u>. Horns are widely used as antennas at UHF and microwave frequencies, above 300 MHz. They are used as **feed antennas** (called feed horns) for larger antenna structures such as parabolic antennas, as standard calibration antennas to measure the gain of other antennas, and as directive antennas for such devices

as radar guns, automatic door openers, and microwave radiometers. Their advantages are moderate directivity, broad bandwidth, low losses, and simple construction and adjustment.

One of the first horn antennas was constructed in 1897 by Bengali-Indian radio researcher Jagadish Chandra Bose in his pioneering experiments with microwaves. The modern horn antenna was invented independently in 1938 by Wilmer Barrow and G. C. Southworth The development of radar in World War II stimulated horn research to design feed horns for radar antennas. The corrugated horn invented by Kay in 1962 has become widely used as a feed horn for microwave antennas such as satellite dishes and radio telescopes.

<u>An advantage of horn antennas is that since they have no resonant elements, they can operate over a wide range of frequencies, a wide bandwidth</u>. The usable bandwidth of horn antennas is typically of the order of 10:1, and can be up to 20:1 (for example allowing it to operate from 1 GHz to 20 GHz). The input impedance is slowly varying over this wide frequency range, allowing low voltage standing wave ratio (VSWR) over the bandwidth. The gain of horn antennas ranges up to 25 dBi, with 10–20 dBi being typical.

I guess the fact that people are called cattle

would just be another coincidence, since these antennas create a feed?

Feed - Old English fedan "nourish, give food to, sustain, foster" (transitive), from Proto-Germanic *fodjan (source also of Old Saxon fodjan, Old Frisian feda, Dutch voeden, Old High German fuotan, Old Norse foeða, Gothic fodjan "to feed"), from PIE root *pa- "to feed." Intransitive sense "take food, eat" is from late 14c. Meaning "to supply to as food" is from 1818.

"action of feeding," 1570s, from feed (v.). Meaning "food for animals" is first attested 1580s. Meaning "a sumptuous meal" is from 1808. Of machinery, "action of or system for providing raw material" from 1892.

Feed - cause to move gradually and steadily, typically through a confined space; a device or conduit for supplying material to a machine. Distribute (a broadcast) to local television or radio stations via satellite or network; a broadcast distributed by a satellite or network from a central source to a large number of radio.

Feed - a line or prompt given to an actor on stage.

I have demonstrated beyond a shadow of a doubt, like no other person has that you are a complex electronic device. Don't you get that yet? You need a signal to be online family, that can come from God, Satan or man acting on behalf of either party. In electrical engineering, a feeder line is a type of transmission line. In addition Feeders are the power lines through which electricity is transmitted in power systems. Feeder transmits power from Generating station or substation to the distribution points. KEEP IN THE FRONT OF YOUR MIND THAT YOUR SPINE IS A FM ANTENNA!!!

Investigation of the spinal cord as a natural receptor antenna for incident electromagnetic waves and possible impact on the central nervous system

Sevaiyan Balaguru 1, Rajan Uppal, Ravinder Pal Vaid, Balasubramaniam Preetham Kumar

Abstract

The effects of electromagnetic field (EMF) exposure on biological systems have been studied for many years, both as a source of medical therapy and also for potential health risks. In particular, the mechanisms of EMF absorption in the human or animal body is of medical/engineering interest, and modern modelling techniques, such as the Finite Difference Time Domain (FDTD), can be utilized to simulate the voltages and currents induced in different parts of the body. The simulation of one particular component, the spinal cord, is the focus of this article, and this study is motivated by the fact that the spinal cord can be modelled as a linear conducting structure, capable of generating a significant amount of voltage from incident EMF. In this article, we show, through a FDTD simulation analysis of an incoming electromagnetic field (EMF), that the spinal cord acts as a natural antenna, with frequency dependent induced electric voltage and current distribution. The multi-frequency (100-2400 MHz) simulation results show that peak voltage and current response is observed in the FM radio range around 100 MHz, with significant strength to potentially cause changes in the CNS. **This work can contribute to the understanding of the mechanism behind EMF energy leakage into the CNS, and the possible contribution of the latter energy leakage towards the weakening of the blood brain barrier (BBB), whose degradation is associated with the progress of many diseases, including Acquired Immuno-Deficiency Syndrome (AIDS)**.

Think about the ramifications, wait **RAM**ifications! In the same way as we discussed the possibility that Autism may exist in this space, how many illnesses especially 'mental' are related to this unknown aspect of our Anatomy!? How much of this is purposeful?

I know this is too scary for you soooooo, you like but Doc, the governments don't know all this, right???

You think that TV, Radio, Missiles, Wifi add infinitum exist and they don't know? Go reread the Electrician Prayer Manual!!! Have you ever heard the name Tesla? You do know Trump is in power via Tesla right? OK... Wait you understand the connection between Elon and Trump is deeper than politics, that's why I am a fan of Trump! You don't understand how much I am changing your life, empowering you! John George Trump (August 21, 1907 –

February 21, 1985) was an American electrical engineer, inventor, and physicist. A professor at the Massachusetts Institute of Technology (MIT) from 1936 to 1985, he was a recipient of the National Medal of Science and a member of the National Academy of Engineering. Trump was noted for developing rotational radiation therapy. Together with Robert J. Van de Graaff, he developed one of the first million-volt X-ray generators. He was the uncle of Donald Trump.

During 1942, Trump became secretary of the microwave committee, a sub-committee of the NDRC. The director of the microwave committee was **Alfred Lee Loomis**, the "millionaire physicist", who decided to create a laboratory. He selected a site for it, chose a suitably discreet and ambiguous name for it, and funded its construction, until governmental administration was established. The new institution came to be called the **MIT Radiation Laboratory**, or the "Rad Lab". As wartime shortages in Britain increased, many of its radar researchers would move to the well-funded laboratory at MIT, where they helped create groundbreaking progress in developing practical devices and systems, which would see widespread field deployment in combat. On January 9, 1943, two days after Nikola Tesla died destitute in a New York City hotel, the FBI called MIT professor and esteemed electrical engineer, John G. Trump, to determine if any of the belongings in the inventor's estate—which included a purported weapon of mass destruction Tesla called the death ray—would be dangerous if they fell into enemy hands. Bruh!!! Why you think Trump is pumping that Bible. He know what time it is! He know what Time Square is Bwahahahahaha....

There are other reasons I created the specific blend of **<u>Algae called Bleu Magick</u>**! You have to keep that in your kitchen!!! You have to keep that in your plasma!!!

BOOK OF HORNS

Horny - late 14c., "made of horn," from horn (n.) + -y (2). From 1690s as "callous, resembling horn." The colloquial meaning "lustful, sexually aroused," was in use certainly by 1889, perhaps as early as 1863; it probably derives from the late 18c. slang expression to have the horn, suggestive of male sexual excitement (but eventually applied to women as well); see horn (n.). As a noun it once also was a popular name for a domestic cow. For an adjective in the original sense of the word, hornish (1630s) and horn-like (1570s) are available.

You think the chakra thing is a game, but Blavatsky and team changed Horny into a sexual term! They turned horny from a divine thing to a lustful thing!!! Wakey wakey!!! Yeah this is another 'Black' thing! Oversexual creatures!!! Super Horny!!! Born Sinners!!! All these floating stories about 'Black' people create 'space' for the stereo types we still live by, this is why no cares we are the leaders of death by Diabetes, Cancer, STDs etc... I do though!!! Eumelanin makes you more resistant to AntiFungal Drugs and more susceptible to Opioids... Wait a minute wait a minute.... Before we even go back into the 'sins of the Flesh'... **Have you figured out why belief or faith is required yet**? Let's just imagine for a second that Dr. EnQi is not insane.... Lmao....

God is Music
Nature is his Producer
Pigment is his beat Machine...

You are saved through Faith, because Faith is basically being tuned into Gods signal! You have free will, the ability to turn the dial. Aren't we taught via Motivational Speakers that the key to changing our '**STATION**' in life is based on how we focus our mind.

Faith - mid-13c., faith, feith, fei, fai "faithfulness to a trust or promise; loyalty to a person; honesty, truthfulness," from Anglo-French and Old French feid, foi "faith, belief, trust, confidence; pledge" (11c.), from Latin fides "trust, faith, confidence, reliance, credence, belief," from root of fidere "to trust,"from PIE root *bheidh- "to trust, confide, persuade." For sense evolution, compare belief. It has been accommodated to other English abstract nouns in -th (truth, health, etc.).

From early 14c. as "assent of the mind to the truth of a statement for which there is incomplete evidence," especially "belief in religious matters" (matched with hope and charity). Since mid-14c. in reference to the Christian church or religion; from late 14c. in reference to any religious persuasion.

And faith is neither the submission of the reason, nor is it the acceptance, simply and absolutely upon testimony, of what reason cannot reach. Faith is: the being able to cleave to a power of goodness appealing to our higher and real self, not to our lower and apparent self.
[Matthew Arnold, "Literature & Dogma," 1873]
From late 14c. as "confidence in a person or thing with reference to truthfulness or reliability," also "fidelity of one spouse to another." Also in Middle English "a sworn oath," hence its frequent use in Middle English oaths and asseverations (par ma fay, mid-13c.; bi my fay, c. 1300).

Station - late 13c., stacioun, "a place one normally occupies," from Old French stacion, estacion "site, location; **station of the Cross**; stop, standstill," from Latin stationem (nominative statio) "a standing, standing firm; a post, job, position; military post; a watch, guard, sentinel;

anchorage, port" (related to stare "to stand," from PIE root *sta- "to stand, make or be firm").

The meaning "fixed uniform distance in surveying" is from 1570s. The meaning "each of a number of holy places visited in succession by pilgrims" is from late 14c. in English; a similar notion is in Stations of the Cross (1550s). The meaning "regular stopping place" is recorded by 1797, in reference to coach routes; it was applied to stopping places on railroads by 1830.

The meaning "military post" in English is from c. 1600. The meaning "place where people are placed or sent for some special purpose, locality to which a functionary is appointed" (as in polling station) is by 1817, in police station "place where a police force is assembled when not on duty;" station house "police station" is attested from 1836.

The meaning "**place for transmitting radio or television signals**" is from **1912**, in radio station; station break, a pause in broadcasting to give the local station a chance to identify itself, is attested from 1942.

The figurative or extended sense of "status, rank" (one's "place" in the scale of society) is from c. 1600.

Dial - early 15c., "sundial, instrument for indicating the hour of the day by means of a shadow thrown upon a graduated surface," earlier "**dial of a compass**" (mid-14c.), from Old French dyal, apparently from Medieval Latin dialis "daily," from Latin dies "day," from PIE root *dyeu- "**to shine**." The word perhaps was abstracted from a phrase such as Medieval Latin rota dialis "daily wheel."

It evolved to mean any round plate or face over which a pointer moves to indicate something about the machinery to which it is attached. Sense of "**face of a clock (or later a watch)**, upon which hours and minutes are marked and over which the hands move" is from mid-15c.
Telephone sense "circular plate marked with numbers and letters which can be rotated to establish connection" is from 1879, which led to dial tone (1921), "the signal to begin dialing." Dial-plate is attested from 1680s.

1650s, "to work with aid of a dial or compass; measure with or as with a dial," from dial (n.). Sense of "**rotate the dial plate of a telephone to indicate the number to which a connection is to be established**" is from 1921. Related: Dialed; dialing.

Matthew 17

And after six days Jesus taketh Peter, James, and John his brother, and bringeth them up into an high mountain apart,
2 And was transfigured before them: and his face did shine as the sun, and his raiment was white

as the light.

3 And, behold, there appeared unto them Moses and Elias talking with him.

4 Then answered Peter, and said unto Jesus, Lord, it is good for us to be here: if thou wilt, let us make here three tabernacles; one for thee, and one for Moses, and one for Elias.

5 While he yet spake, behold, a bright cloud overshadowed them: and behold a voice out of the cloud, which said, This is my beloved Son, in whom I am well pleased; hear ye him.

6 And when the disciples heard it, they fell on their face, and were sore afraid.

7 And Jesus came and touched them, and said, Arise, and be not afraid.

8 And when they had lifted up their eyes, they saw no man, save Jesus only.

9 And as they came down from the mountain, Jesus charged them, saying, Tell the vision to no man, until the Son of man be risen again from the dead.

10 And his disciples asked him, saying, Why then say the scribes that Elias must first come?

11 And Jesus answered and said unto them, Elias truly shall first come, and restore all things.

12 But I say unto you, That Elias is come already, and they knew him not, but have done unto him whatsoever they listed. Likewise shall also the Son of man suffer of them.

13 Then the disciples understood that he spake unto them of John the Baptist.

14 And when they were come to the multitude, there came to him a certain man, kneeling down to him, and saying,

15 Lord, have mercy on my son: for he is lunatick, and sore vexed: for ofttimes he falleth into the fire, and oft into the water.

16 And I brought him to thy disciples, and they could not cure him.

17 Then Jesus answered and said, O faithless and perverse generation, how long shall I be with you? how long shall I suffer you? bring him hither to me.

18 And Jesus rebuked the devil; and he departed out of him: and the child was cured from that very hour.

19 Then came the disciples to Jesus apart, and said, Why could not we cast him out?

20 **And Jesus said unto them, Because of your unbelief: for verily I say unto you, If ye have faith as a grain of mustard seed, ye shall say unto this mountain, Remove hence to yonder place; and it shall remove; and nothing shall be impossible unto you.**

21 Howbeit this kind goeth not out but by prayer and fasting.

22 And while they abode in Galilee, Jesus said unto them, The Son of man shall be betrayed into the hands of men:

23 And they shall kill him, and the third day he shall be raised again. And they were exceeding sorry.

24 And when they were come to Capernaum, they that received tribute money came to Peter, and said, Doth not your master pay tribute?

25 He saith, Yes. And when he was come into the house, Jesus prevented him, saying, What thinkest thou, Simon? of whom do the kings of the earth take custom or tribute? of their own children, or of strangers?

26 Peter saith unto him, Of strangers. Jesus saith unto him, Then are the children free.

27 Notwithstanding, lest we should offend them, go thou to the sea, and cast an hook, and take up the fish that first cometh up; and when thou hast opened his mouth, thou shalt find a piece of money: that take, and give unto them for me and thee.

You know as a Certified Master Herbalist I found out that, mustard plants can absorb the widest range of metals, even gold! Just in case you don't know, most metals can pick up radio waves and convert the photons into electrons via the photoelectric effect, causing them to vibrate. This means, what regarding all them metals your body requires for Health, the metals that disrupt your Health?

OK... Back to work...

Radiate - "having rays, furnished with rays or ray-like parts, shining," 1660s, from Latin radiatus, past participle of radiare "to beam, shine, gleam; make beaming," from radius "beam of light; spoke of a wheel" (see radius).

1610s, "issue or spread in all directions from a point in rays or straight lines," from Latin radiatus, past participle of radiare "to beam, shine, gleam; make beaming," from radius "beam of light; spoke of a wheel" (see radius). Meaning "be radiant, give off rays (of light or heat)" is from 1640s. Related: Radiated; radiates; radiating.

Radiator - 1832, "any thing which radiates," agent noun in Latin form from radiate (v.). Originally a stove-like apparatus, as a device designed to communicate heat from steam to a room by 1855; the sense of "cooling device in an internal combustion engine" is by 1899.

Radiation - mid-15c., radiacion, "act or process of emitting light," from Latin radiationem (nominative radiatio) "a shining, radiation," noun of action from past-participle stem of radiare "to beam, shine, gleam; make beaming," from radius "beam of light; spoke of a wheel" (see radius).

Meaning "rays or beams emitted" is from 1560s. Meaning "divergence from a center" is 1650s. In modern physics, "emission or transmission of energy in the form of waves or particles," especially in reference to ionizing radiation, from early 20c.

Pigment - late 14c., "a red dye," from Latin pigmentum "coloring matter, pigment, paint," figuratively "ornament," from stem of pingere "to color, paint" (see paint (v.)). By 1610s in the broader sense "**any substance that is or can be used by painters to impart color**" (technically a dry substance that can be powdered and mixed with a liquid medium).

Variants of this word could have been known in Old English and Middle English (compare 12c. pyhmentum, later piment) with a sense of "a spiced drink, a remedy or concoction containing spices," based on a secondary sense of the Latin word in Medieval Latin. As a verb from 1900. Related: Pigmented. Also pigmental "of or pertaining
to pigment" (1836); pigmentary (1835).

Melanin - dark brown or black pigment found in animal bodies, 1832, Modern Latin, with chemical suffix -in (2); the first element is from Greek melas (genitive melanos) "black" (see melano-). Related: Melanism; melanistic.

Melano - word-forming element meaning "black," from Greek melano-, combining form of melas (genitive melanos) "black, dark, murky,"probably from a PIE root *melh-"black, of darkish color" (source also of

Sanskrit malinah "dirty, stained, black," Lithuanian mėlynas "blue," Latin mulleus "reddish").

Eu - word-forming element, in modern use meaning "good, well," from Greek eus "**good**," eu "**well**" (adv.), also "luckily, happily" (opposed to kakos), as a noun, "**the right**, the good cause," from PIE *(e)su- "good" (source also of Sanskrit su- "good," Avestan hu- "good"), originally a suffixed form of root *es- "to be." In compounds the Greek word had more a sense of "greatness, **abundance**, **prosperity**," and was opposed to dys-.

Proto-Indo-European root meaning "**to dress**," with extended form *wes- (2) "**to clothe**."
It is the hypothetical source of/evidence for its existence is provided by: Hittite washshush "**garments**," washanzi "they dress;" Sanskrit vaste "he puts on," vasanam "garment;" Avestan vah-; Greek esthes "clothing," hennymi "to clothe," eima "garment;" Latin vestire "to clothe;" Welsh gwisgo, Breton gwiska; Old English werian "to clothe, put on, cover up," wæstling "sheet, blanket."

Would God give his children the wrong clothing? Have you read the Bible??? God is Light, your body is his Temple, your Flesh is the Word... You have nothing to be ashamed about, God gave you abundance and prosperity of the Spirit!!! Science gives you two 2s but they don't 2+2=4 for you. Science tells you the Electromagnetism comes from the Cross, they tell Radio waves are the beginning, they tell you life is electromagnetic etc.... German physicist Heinrich Hertz (is credited as the person who) discovered radio waves, a milestone widely seen as confirmation of James Clerk Maxwell's electromagnetic theory and which paved the way for numerous advances in communication technology. Born in Hamburg on February 22, 1857, Hertz was the eldest of five children. The wavelengths of radio waves range from thousands of metres to 30 cm. These correspond to frequencies as low as 3 Hz and as high as 1 gigahertz (10 to the 9th power Hz). Wait the Most High Radio waves are 10 to the 9th power, if you see it, you see it...

In the back of this book you will find a Article providing your proof that EuMelanin is absolutely active with Radio Waves.

Genesis 3

1 Now the serpent was more subtil than any beast of the field which the Lord God had made. And he said unto the woman, Yea, hath God said, Ye shall not eat of every tree of the garden?

2 And the woman said unto the serpent, We may eat of the fruit of the trees of the garden:

3 But of the fruit of the tree which is in the midst

of the garden, God hath said, Ye shall not eat of it, neither shall ye touch it, lest ye die.

4 And the serpent said unto the woman, Ye shall not surely die:

5 For God doth know that in the day ye eat thereof, then your eyes shall be opened, and ye shall be as gods, knowing good and evil.

6 And when the woman saw that the tree was good for food, and that it was pleasant to the eyes, and a tree to be desired to make one wise, she took of the fruit thereof, and did eat, and gave also unto her husband with her; and he did eat.

7 And the eyes of them both were opened, and they knew that they were naked; and they sewed fig leaves together, and made themselves aprons.

8 And they heard the voice of the Lord God walking in the garden in the cool of the day: and Adam and his wife hid themselves from the presence of the Lord God amongst the trees of the garden.

9 And the Lord God called unto Adam, and said unto him, Where art thou?

10 And he said, I heard thy voice in the garden, and I was afraid, because I was naked; and I hid myself.

11 And he said, Who told thee that thou wast naked? Hast thou eaten of the tree, whereof I commanded thee that thou shouldest not eat?

12 And the man said, The woman whom thou gavest to be with me, she gave me of the tree, and I did eat.

13 And the Lord God said unto the woman, What is this that thou hast done? And the woman said, The serpent beguiled me, and I did eat.

14 And the Lord God said unto the serpent, Because thou hast done this, thou art cursed above all cattle, and above every beast of the field; upon thy belly shalt thou go, and dust shalt thou eat all the days of thy life:

15 And I will put enmity between thee and the woman, and between thy seed and her seed; it shall bruise thy head, and thou shalt bruise his heel.

16 Unto the woman he said, I will greatly multiply thy sorrow and thy conception; in sorrow thou shalt bring forth children; and thy desire shall be to thy husband, and he shall rule over thee.

17 And unto Adam he said, Because thou hast hearkened unto the voice of thy wife, and hast eaten of the tree, of which I commanded thee, saying, Thou shalt not eat of it: cursed is the

ground for thy sake; in sorrow shalt thou eat of it all the days of thy life;

18 Thorns also and thistles shall it bring forth to thee; and thou shalt eat the herb of the field;

19 In the sweat of thy face shalt thou eat bread, till thou return unto the ground; for out of it wast thou taken: for dust thou art, and unto dust shalt thou return.

20 And Adam called his wife's name Eve; because she was the mother of all living.

21 **<u>Unto Adam also and to his wife did the Lord God make coats of skins, and clothed them</u>**.

22 **<u>And the Lord God said, Behold, the man is become as one of us, to know good and evil</u>**: and now, lest he put forth his hand, and take also of the tree of life, and eat, and live for ever:

23 Therefore the Lord God sent him forth from the garden of Eden, to till the ground from whence he was taken.

24 So he drove out the man; and he placed at the east of the garden of Eden Cherubims, and a flaming sword which turned every way, to keep the way of the tree of life.

Interpret it how you want, but man became like

God after receiving his skin. Please turn back to the definitions we just did. Add that to the fact that god punishes through the skin as well. It makes sense if the word is flesh, and through it you are blessed. I take away my children's electronics all the time! God was simply taking away his children's electronics!!!

DOC, HORNS, MAKE THIS CHAPTER MAKE SENSE PLEASE? Such good questions. If you ask around or search around, most sources will slap you down so hard and fast that you stop looking. Here is the slickness though, the sweet spot. An RF signal can have the same frequency as a sound wave, and most people can hear a 5 kHz audio tone. No one can hear a 5 kHz RF signal. The audio tone is compression waves traveling through air that your ears can pick up. The RF signal is waves in the electromagnetic field that you ears have no way of picking up, your skin can feel those low frequencies though. Your phone runs on RF waves, you feel it heat up all the time right? LMAO... I know its not the same but you get what I am saying (I hope). Your Spine can catch the FM wavelengths, we convert infrared into Heat & Melatonin all the time. This is why it is important that we understand melanin vs pigment. Our flesh has way more pigments than Melanin!!! We have many books on them particularly the Autodidact & Orthorexia books, please read &/or reread them carefully!!!

Let add this difference between AM & FM: When comparing AM radio waves with FM radio waves, the former, which operates in the kilohertz range, has a longer wavelength compared to FM radio waves, which operate in the megahertz range.

This is because frequency and wavelength have an inverse relationship, as described by the formula $c = \lambda \nu$, where c is the speed of light, λ (lambda) is the wavelength, and ν (nu) is the frequency. Therefore, lower frequency (such as AM radio) corresponds to a longer wavelength compared to higher frequency (FM radio) waves. The test in the back of the book were mostly on AM but I have included the spine info as well... Your skin picks up AM and the Spine FM...

Here I think is a good place to mention real dancing vs choreography, Choreography is NOT dancing!!! That is some Mirror Neuron activity, but the Bible speaks on this...

Isaiah 42

1 Behold my servant, whom I uphold; mine elect, in whom my soul delighteth; I have put my spirit upon him: he shall bring forth judgment to the Gentiles.

2 He shall not cry, nor lift up, nor cause his voice to be heard in the street.

3 A bruised reed shall he not break, and the smoking flax shall he not quench: he shall bring forth judgment unto truth.

4 He shall not fail nor be discouraged, till he have set judgment in the earth: and the isles shall wait for his law.

5 Thus saith God the Lord, he that created the heavens, and stretched them out; he that spread forth the earth, and that which cometh out of it; he that giveth breath unto the people upon it, and **spirit to them that walk therein**:

6 **I the Lord have called thee in righteousness, and will hold thine hand, and will keep thee, and give thee for a covenant of the people, for a light of the Gentiles**;

7 To open the blind eyes, to bring out the prisoners from the prison, and them that sit in darkness out of the prison house.

8 I am the Lord: that is my name: and my glory will I not give to another, neither my praise to graven images.

9 Behold, the former things are come to pass, and new things do I declare: before they spring forth I tell you of them.

10 Sing unto the Lord a new song, and his praise from the end of the earth, ye that go down to the sea, and all that is therein; the isles, and the inhabitants thereof.

11 Let the wilderness and the cities thereof lift up their voice, the villages that Kedar doth inhabit: let the inhabitants of the rock sing, let them shout from the top of the mountains.

12 Let them give glory unto the Lord, and declare his praise in the islands.

13 The Lord shall go forth as a mighty man, he shall stir up jealousy like a man of war: he shall cry, yea, roar; he shall prevail against his enemies.

14 I have long time holden my peace; I have been still, and refrained myself: now will I cry like a travailing woman; I will destroy and devour at once.

15 I will make waste mountains and hills, and dry up all their herbs; and I will make the rivers islands, and I will dry up the pools.

16 And I will bring the blind by a way that they knew not; I will lead them in paths that they have not known: I will make darkness light before them, and crooked things straight. These things will I do unto them, and not forsake them.

17 They shall be turned back, they shall be greatly ashamed, that trust in graven images, that say to the molten images, Ye are our gods.

18 Hear, ye deaf; and look, ye blind, that ye may see.

19 Who is blind, but my servant? or deaf, as my messenger that I sent? who is blind as he that is perfect, and blind as the Lord's servant?

20 Seeing many things, but thou observest not; opening the ears, but he heareth not.

21 The Lord is well pleased for his righteousness' sake; he will magnify the law, and make it honourable.

22 But this is a people robbed and spoiled; they are all of them snared in holes, and they are hid

in prison houses: they are for a prey, and none delivereth; for a spoil, and none saith, Restore.

23 Who among you will give ear to this? who will hearken and hear for the time to come?

24 Who gave Jacob for a spoil, and Israel to the robbers? did not the Lord, he against whom we have sinned? for they would not walk in his ways, neither were they obedient unto his law.

25 Therefore he hath poured upon him the fury of his anger, and the strength of battle: and it hath set him on fire round about, yet he knew not; and it burned him, yet he laid it not to heart.

Radio waves are a type of electromagnetic radiation with the lowest frequencies and the longest wavelengths in the electromagnetic spectrum, typically (meaning usually but not the complete scale) with frequencies below 100 gigahertz (GHz) and wavelengths greater than 1 millimeter ($3/64$ inch), about the diameter of a grain of rice. Radio waves with frequencies above about 1 GHz and wavelengths shorter than 30 centimeters are called microwaves. Like all electromagnetic waves, radio waves in vacuum travel at the speed of light, and in the Earth's atmosphere at a slightly lower speed. Radio waves are generated by charged particles undergoing acceleration, such as time-varying electric currents. Naturally occurring radio waves are emitted by lightning and astronomical objects, and are part of the blackbody radiation emitted by all warm objects.Radio waves have the longest wavelengths in the electromagnetic spectrum. They range from the length of a football to larger than our planet. The US military has been using extremely low frequency (ELF) radio waves from 3 Hz to 30 Hz to communicate with submarines. ELF waves can penetrate seawater, therefore the US and Russian militaries have used ELF transmission facilities to communicate with their deep submerged submarines. ELF is important from a public health standpoint because of the widespread use of electrical current at the 50 or 60 Hz frequency.

Sound Range: 20 to 20,000 Hz

Voice Range 90 to 255 Hz

Radio Range: 1 hertz up to 3,000 billion hertz.

Infrared Range 1 Trillion Hertz to ... this is where our body heat is...

Get it? Sound Waves are Radio Waves within a medium, still not exactly but interesting. Right? Light waves come from the movement at the subatomic size scale and sound waves come from movement in the matter scale. As above so below, the below creates the above though.

The use of horns aka antennas is all throughout the Bible, I fear people won't take this seriously though. There are different types of Horns in the

Bible it's a book of horns, that further confuses people.

Joshua 6:4-5, 20

Now Jericho was straitly shut up because of the children of Israel: none went out, and none came in.

2 And the Lord said unto Joshua, See, I have given into thine hand Jericho, and the king thereof, and the mighty men of valour.

3 And ye shall compass the city, all ye men of war, and go round about the city once. Thus shalt thou do six days.

4 And seven priests shall bear before the ark seven trumpets of rams' horns: and the seventh day ye shall compass the city seven times, and the priests shall blow with the trumpets.

5 And it shall come to pass, that when they make a long blast with the ram's horn, and when ye hear the sound of the trumpet, all the people shall shout with a great shout; and the wall of the city shall fall down flat, and the people shall ascend up every man straight before him.

6 And Joshua the son of Nun called the priests, and said unto them, Take up the ark of the covenant, and let seven priests bear seven trumpets of rams' horns before the ark of

the Lord.

7 And he said unto the people, Pass on, and compass the city, and let him that is armed pass on before the ark of the Lord.

8 And it came to pass, when Joshua had spoken unto the people, that the seven priests bearing the seven trumpets of rams' horns passed on before the Lord, and blew with the trumpets: and the ark of the covenant of the Lord followed them.

9 And the armed men went before the priests that blew with the trumpets, and the rereward came after the ark, the priests going on, and blowing with the trumpets.

10 And Joshua had commanded the people, saying, Ye shall not shout, nor make any noise with your voice, neither shall any word proceed out of your mouth, until the day I bid you shout; then shall ye shout.

11 So the ark of the Lord compassed the city, going about it once: and they came into the camp, and lodged in the camp.

12 And Joshua rose early in the morning, and the priests took up the ark of the Lord.

13 And seven priests bearing seven trumpets of rams' horns before the ark of the Lord went on continually, and blew with the trumpets: and the armed men went before them; but the rereward came after the ark of the Lord, the priests going on, and blowing with the trumpets.

14 And the second day they compassed the city

once, and returned into the camp: so they did six days.

15 And it came to pass on the seventh day, that they rose early about the dawning of the day, and compassed the city after the same manner seven times: only on that day they compassed the city seven times.

16 And it came to pass at the seventh time, when the priests blew with the trumpets, Joshua said unto the people, Shout; for the Lord hath given you the city.

17 And the city shall be accursed, even it, and all that are therein, to the Lord: only Rahab the harlot shall live, she and all that are with her in the house, because she hid the messengers that we sent.

18 And ye, in any wise keep yourselves from the accursed thing, lest ye make yourselves accursed, when ye take of the accursed thing, and make the camp of Israel a curse, and trouble it.

19 But all the silver, and gold, and vessels of brass and iron, are consecrated unto the Lord: they shall come into the treasury of the Lord.

20 So the people shouted when the priests blew with the trumpets: and it came to pass, when the people heard the sound of the trumpet, and the people shouted with a great shout, that the wall fell down flat, so that the people went up into the city, every man straight before him, and they took the city.

21 And they utterly destroyed all that was in

the city, both man and woman, young and old, and ox, and sheep, and ass, with the edge of the sword.

22 But Joshua had said unto the two men that had spied out the country, Go into the harlot's house, and bring out thence the woman, and all that she hath, as ye sware unto her.

23 And the young men that were spies went in, and brought out Rahab, and her father, and her mother, and her brethren, and all that she had; and they brought out all her kindred, and left them without the camp of Israel.

24 And they burnt the city with fire, and all that was therein: only the silver, and the gold, and the vessels of brass and of iron, they put into the treasury of the house of the Lord.

25 And Joshua saved Rahab the harlot alive, and her father's household, and all that she had; and she dwelleth in Israel even unto this day; because she hid the messengers, which Joshua sent to spy out Jericho.

26 And Joshua adjured them at that time, saying, Cursed be the man before the Lord, that riseth up and buildeth this city Jericho: he shall lay the foundation thereof in his firstborn, and in his youngest son shall he set up the gates of it.

27 So the Lord was with Joshua; and his fame was noised throughout all the country.

Numbers 10:2

1 And the Lord spake unto Moses, saying,

2 Make thee two trumpets of silver; of a whole piece shalt thou make them: that thou mayest use them for the calling of the assembly, and for the journeying of the camps.

3 And when they shall blow with them, all the assembly shall assemble themselves to thee at the door of the tabernacle of the congregation.

4 And if they blow but with one trumpet, then the princes, which are heads of the thousands of Israel, shall gather themselves unto thee.

5 When ye blow an alarm, then the camps that lie on the east parts shall go forward.

6 When ye blow an alarm the second time, then the camps that lie on the south side shall take their journey: they shall blow an alarm for their journeys.

7 But when the congregation is to be gathered together, ye shall blow, but ye shall not sound an alarm.

8 And the sons of Aaron, the priests, shall blow with the trumpets; and they shall be to you for an ordinance for ever throughout your generations.

9 And if ye go to war in your land against the enemy that oppresseth you, then ye shall blow an alarm with the trumpets; and ye shall be remembered before the Lord your God, and ye shall be saved from your enemies.

10 Also in the day of your gladness, and in

your solemn days, and in the beginnings of your months, ye shall blow with the trumpets over your burnt offerings, and over the sacrifices of your peace offerings; that they may be to you for a memorial before your God: I am the Lord your God.

11 And it came to pass on the twentieth day of the second month, in the second year, that the cloud was taken up from off the tabernacle of the testimony.

12 And the children of Israel took their journeys out of the wilderness of Sinai; and the cloud rested in the wilderness of Paran.

13 And they first took their journey according to the commandment of the Lord by the hand of Moses.

14 In the first place went the standard of the camp of the children of Judah according to their armies: and over his host was Nahshon the son of Amminadab.

15 And over the host of the tribe of the children of Issachar was Nethaneel the son of Zuar.

16 And over the host of the tribe of the children of Zebulun was Eliab the son of Helon.

17 And the tabernacle was taken down; and the sons of Gershon and the sons of Merari set forward, bearing the tabernacle.

18 And the standard of the camp of Reuben set forward according to their armies: and over his host was Elizur the son of Shedeur.

19 And over the host of the tribe of the children

of Simeon was Shelumiel the son of Zurishaddai.

20 And over the host of the tribe of the children of Gad was Eliasaph the son of Deuel.

21 And the Kohathites set forward, bearing the sanctuary: and the other did set up the tabernacle against they came.

22 And the standard of the camp of the children of Ephraim set forward according to their armies: and over his host was Elishama the son of Ammihud.

23 And over the host of the tribe of the children of Manasseh was Gamaliel the son of Pedahzur.

24 And over the host of the tribe of the children of Benjamin was Abidan the son of Gideoni.

25 And the standard of the camp of the children of Dan set forward, which was the rereward of all the camps throughout their hosts: and over his host was Ahiezer the son of Ammishaddai.

26 And over the host of the tribe of the children of Asher was Pagiel the son of Ocran.

27 And over the host of the tribe of the children of Naphtali was Ahira the son of Enan.

28 Thus were the journeyings of the children of Israel according to their armies, when they set forward.

29 And Moses said unto Hobab, the son of Raguel the Midianite, Moses' father in law, We are journeying unto the place of which the Lord said, I will give it you: come thou with us, and we will do thee good: for the Lord hath spoken good concerning Israel.

30 And he said unto him, I will not go; but I will depart to mine own land, and to my kindred.

31 And he said, Leave us not, I pray thee; forasmuch as thou knowest how we are to encamp in the wilderness, and thou mayest be to us instead of eyes.

32 And it shall be, if thou go with us, yea, it shall be, that what goodness the Lord shall do unto us, the same will we do unto thee.

33 And they departed from the mount of the Lord three days' journey: and the ark of the covenant of the Lord went before them in the three days' journey, to search out a resting place for them.

34 And the cloud of the Lord was upon them by day, when they went out of the camp.

35 And it came to pass, when the ark set forward, that Moses said, Rise up, Lord, and let thine enemies be scattered; and let them that hate thee flee before thee.

36 And when it rested, he said, Return, O Lord, unto the many thousands of Israel.

Exodus 10

And thou shalt make an altar to burn incense upon: of shittim wood shalt thou make it.

2 A cubit shall be the length thereof, and a cubit the breadth thereof; foursquare shall it be: and two cubits shall be the height thereof: the horns thereof shall be of the same.

3 And thou shalt overlay it with pure gold, the

top thereof, and the sides thereof round about, and the horns thereof; and thou shalt make unto it a crown of gold round about.

4 And two golden rings shalt thou make to it under the crown of it, by the two corners thereof, upon the two sides of it shalt thou make it; and they shall be for places for the staves to bear it withal.

5 And thou shalt make the staves of shittim wood, and overlay them with gold.

6 And thou shalt put it before the vail that is by the ark of the testimony, before the mercy seat that is over the testimony, where I will meet with thee.

7 And Aaron shall burn thereon sweet incense every morning: when he dresseth the lamps, he shall burn incense upon it.

8 And when Aaron lighteth the lamps at even, he shall burn incense upon it, a perpetual incense before the Lord throughout your generations.

9 Ye shall offer no strange incense thereon, nor burnt sacrifice, nor meat offering; neither shall ye pour drink offering thereon.

10 And Aaron shall make an atonement upon the horns of it once in a year with the blood of the sin offering of atonements: once in the year shall he make atonement upon it throughout your generations: it is most holy unto the Lord.

11 And the Lord spake unto Moses, saying,

12 When thou takest the sum of the children of Israel after their number, then shall they

give every man a ransom for his soul unto the Lord, when thou numberest them; that there be no plague among them, when thou numberest them.

13 This they shall give, every one that passeth among them that are numbered, half a shekel after the shekel of the sanctuary: (a shekel is twenty gerahs:) an half shekel shall be the offering of the Lord.

14 Every one that passeth among them that are numbered, from twenty years old and above, shall give an offering unto the Lord.

15 The rich shall not give more, and the poor shall not give less than half a shekel, when they give an offering unto the Lord, to make an atonement for your souls.

16 And thou shalt take the atonement money of the children of Israel, and shalt appoint it for the service of the tabernacle of the congregation; that it may be a memorial unto the children of Israel before the Lord, to make an atonement for your souls.

17 And the Lord spake unto Moses, saying,

18 Thou shalt also make a laver of brass, and his foot also of brass, to wash withal: and thou shalt put it between the tabernacle of the congregation and the altar, and thou shalt put water therein.

19 For Aaron and his sons shall wash their hands and their feet thereat:

20 When they go into the tabernacle of the

congregation, they shall wash with water, that they die not; or when they come near to the altar to minister, to burn offering made by fire unto the Lord:

21 So they shall wash their hands and their feet, that they die not: and it shall be a statute for ever to them, even to him and to his seed throughout their generations.

22 Moreover the Lord spake unto Moses, saying,

23 Take thou also unto thee principal spices, of pure myrrh five hundred shekels, and of sweet cinnamon half so much, even two hundred and fifty shekels, and of sweet calamus two hundred and fifty shekels,

24 And of cassia five hundred shekels, after the shekel of the sanctuary, and of oil olive an hin:

25 And thou shalt make it an oil of holy ointment, an ointment compound after the art of the apothecary: it shall be an holy anointing oil.

26 And thou shalt anoint the tabernacle of the congregation therewith, and the ark of the testimony,

27 And the table and all his vessels, and the candlestick and his vessels, and the altar of incense,

28 And the altar of burnt offering with all his vessels, and the laver and his foot.

29 And thou shalt sanctify them, that they may be most holy: whatsoever toucheth them shall be holy.

30 And thou shalt anoint Aaron and his sons, and consecrate them, that they may minister unto me in the priest's office.

31 And thou shalt speak unto the children of Israel, saying, This shall be an holy anointing oil unto me throughout your generations.

32 Upon man's flesh shall it not be poured, neither shall ye make any other like it, after the composition of it: it is holy, and it shall be holy unto you.

33 Whosoever compoundeth any like it, or whosoever putteth any of it upon a stranger, shall even be cut off from his people.

34 And the Lord said unto Moses, Take unto thee sweet spices, stacte, and onycha, and galbanum; these sweet spices with pure frankincense: of each shall there be a like weight:

35 And thou shalt make it a perfume, a confection after the art of the apothecary, tempered together, pure and holy:

36 And thou shalt beat some of it very small, and put of it before the testimony in the tabernacle of the congregation, where I will meet with thee: it shall be unto you most holy.

37 And as for the perfume which thou shalt make, ye shall not make to yourselves according to the composition thereof: it shall be unto thee holy for the Lord.

38 Whosoever shall make like unto that, to smell thereto, shall even be cut off from his people.

Leviticus 4

And the Lord spake unto Moses, saying,

2 Speak unto the children of Israel, saying, If a soul shall sin through ignorance against any of the commandments of the Lord concerning things which ought not to be done, and shall do against any of them:

3 If the priest that is anointed do sin according to the sin of the people; then let him bring for his sin, which he hath sinned, a young bullock without blemish unto the Lord for a sin offering.

4 And he shall bring the bullock unto the door of the tabernacle of the congregation before the Lord; and shall lay his hand upon the bullock's head, and kill the bullock before the Lord.

5 And the priest that is anointed shall take of the bullock's blood, and bring it to the tabernacle of the congregation:

6 And the priest shall dip his finger in the blood, and sprinkle of the blood seven times before the Lord, before the vail of the sanctuary.

7 And the priest shall put some of the blood upon the horns of the altar of sweet incense before the Lord, which is in the tabernacle of the congregation; and shall pour all the blood of the bullock at the bottom of the altar of the burnt offering, which is at the door of the tabernacle of the congregation.

8 And he shall take off from it all the fat of the bullock for the sin offering; the fat that covereth the inwards, and all the fat that is upon the

inwards,

9 And the two kidneys, and the fat that is upon them, which is by the flanks, and the caul above the liver, with the kidneys, it shall he take away,

10 As it was taken off from the bullock of the sacrifice of peace offerings: and the priest shall burn them upon the altar of the burnt offering.

11 And the skin of the bullock, and all his flesh, with his head, and with his legs, and his inwards, and his dung,

12 Even the whole bullock shall he carry forth without the camp unto a clean place, where the ashes are poured out, and burn him on the wood with fire: where the ashes are poured out shall he be burnt.

13 And if the whole congregation of Israel sin through ignorance, and the thing be hid from the eyes of the assembly, and they have done somewhat against any of the commandments of the Lord concerning things which should not be done, and are guilty;

14 When the sin, which they have sinned against it, is known, then the congregation shall offer a young bullock for the sin, and bring him before the tabernacle of the congregation.

15 And the elders of the congregation shall lay their hands upon the head of the bullock before the Lord: and the bullock shall be killed before the Lord.

16 And the priest that is anointed shall bring of the bullock's blood to the tabernacle of the

congregation:

17 And the priest shall dip his finger in some of the blood, and sprinkle it seven times before the Lord, even before the vail.

18 And he shall put some of the blood upon the horns of the altar which is before the Lord, that is in the tabernacle of the congregation, and shall pour out all the blood at the bottom of the altar of the burnt offering, which is at the door of the tabernacle of the congregation.

19 And he shall take all his fat from him, and burn it upon the altar.

20 And he shall do with the bullock as he did with the bullock for a sin offering, so shall he do with this: and the priest shall make an atonement for them, and it shall be forgiven them.

21 And he shall carry forth the bullock without the camp, and burn him as he burned the first bullock: it is a sin offering for the congregation.

22 When a ruler hath sinned, and done somewhat through ignorance against any of the commandments of the Lord his God concerning things which should not be done, and is guilty;

23 Or if his sin, wherein he hath sinned, come to his knowledge; he shall bring his offering, a kid of the goats, a male without blemish:

24 And he shall lay his hand upon the head of the goat, and kill it in the place where they kill the burnt offering before the Lord: it is a sin offering.

25 And the priest shall take of the blood of the

sin offering with his finger, and put it upon the horns of the altar of burnt offering, and shall pour out his blood at the bottom of the altar of burnt offering.

26 And he shall burn all his fat upon the altar, as the fat of the sacrifice of peace offerings: and the priest shall make an atonement for him as concerning his sin, and it shall be forgiven him.

27 And if any one of the common people sin through ignorance, while he doeth somewhat against any of the commandments of the Lord concerning things which ought not to be done, and be guilty;

28 Or if his sin, which he hath sinned, come to his knowledge: then he shall bring his offering, a kid of the goats, a female without blemish, for his sin which he hath sinned.

29 And he shall lay his hand upon the head of the sin offering, and slay the sin offering in the place of the burnt offering.

30 And the priest shall take of the blood thereof with his finger, and put it upon the horns of the altar of burnt offering, and shall pour out all the blood thereof at the bottom of the altar.

31 And he shall take away all the fat thereof, as the fat is taken away from off the sacrifice of peace offerings; and the priest shall burn it upon the altar for a sweet savour unto the Lord; and the priest shall make an atonement for him, and it shall be forgiven him.

32 And if he bring a lamb for a sin offering, he

shall bring it a female without blemish.

33 And he shall lay his hand upon the head of the sin offering, and slay it for a sin offering in the place where they kill the burnt offering.

34 And the priest shall take of the blood of the sin offering with his finger, and put it upon the horns of the altar of burnt offering, and shall pour out all the blood thereof at the bottom of the altar:

35 And he shall take away all the fat thereof, as the fat of the lamb is taken away from the sacrifice of the peace offerings; and the priest shall burn them upon the altar, according to the offerings made by fire unto the Lord: and the priest shall make an atonement for his sin that he hath committed, and it shall be forgiven him.

Leviticus 8

And the Lord spake unto Moses, saying,

2 Take Aaron and his sons with him, and the garments, and the anointing oil, and a bullock for the sin offering, and two rams, and a basket of unleavened bread;

3 And gather thou all the congregation together unto the door of the tabernacle of the congregation.

4 And Moses did as the Lord commanded him; and the assembly was gathered together unto the door of the tabernacle of the congregation.

5 And Moses said unto the congregation, This is the thing which the Lord commanded to be done.

6 And Moses brought Aaron and his sons, and washed them with water.

7 And he put upon him the coat, and girded him with the girdle, and clothed him with the robe, and put the ephod upon him, and he girded him with the curious girdle of the ephod, and bound it unto him therewith.

8 And he put the breastplate upon him: also he put in the breastplate the Urim and the Thummim.

9 And he put the mitre upon his head; also upon the mitre, even upon his forefront, did he put the golden plate, the holy crown; as the Lord commanded Moses.

10 And Moses took the anointing oil, and anointed the tabernacle and all that was therein, and sanctified them.

11 And he sprinkled thereof upon the altar seven times, and anointed the altar and all his vessels, both the laver and his foot, to sanctify them.

12 And he poured of the anointing oil upon Aaron's head, and anointed him, to sanctify him.

13 And Moses brought Aaron's sons, and put coats upon them, and girded them with girdles, and put bonnets upon them; as the Lord commanded Moses.

14 And he brought the bullock for the sin offering: and Aaron and his sons laid their hands upon the head of the bullock for the sin offering.

15 And he slew it; and Moses took the blood, and put it upon the horns of the altar round

about with his finger, and purified the altar, and poured the blood at the bottom of the altar, and sanctified it, to make reconciliation upon it.

16 And he took all the fat that was upon the inwards, and the caul above the liver, and the two kidneys, and their fat, and Moses burned it upon the altar.

17 But the bullock, and his hide, his flesh, and his dung, he burnt with fire without the camp; as the Lord commanded Moses.

18 And he brought the ram for the burnt offering: and Aaron and his sons laid their hands upon the head of the ram.

19 And he killed it; and Moses sprinkled the blood upon the altar round about.

20 And he cut the ram into pieces; and Moses burnt the head, and the pieces, and the fat.

21 And he washed the inwards and the legs in water; and Moses burnt the whole ram upon the altar: it was a burnt sacrifice for a sweet savour, and an offering made by fire unto the Lord; as the Lord commanded Moses.

22 And he brought the other ram, the ram of consecration: and Aaron and his sons laid their hands upon the head of the ram.

23 And he slew it; and Moses took of the blood of it, and put it upon the tip of Aaron's right ear, and upon the thumb of his right hand, and upon the great toe of his right foot.

24 And he brought Aaron's sons, and Moses put of the blood upon the tip of their right ear,

and upon the thumbs of their right hands, and upon the great toes of their right feet: and Moses sprinkled the blood upon the altar round about.

25 And he took the fat, and the rump, and all the fat that was upon the inwards, and the caul above the liver, and the two kidneys, and their fat, and the right shoulder:

26 And out of the basket of unleavened bread, that was before the Lord, he took one unleavened cake, and a cake of oiled bread, and one wafer, and put them on the fat, and upon the right shoulder:

27 And he put all upon Aaron's hands, and upon his sons' hands, and waved them for a wave offering before the Lord.

28 And Moses took them from off their hands, and burnt them on the altar upon the burnt offering: they were consecrations for a sweet savour: it is an offering made by fire unto the Lord.

29 And Moses took the breast, and waved it for a wave offering before the Lord: for of the ram of consecration it was Moses' part; as the Lord commanded Moses.

30 And Moses took of the anointing oil, and of the blood which was upon the altar, and sprinkled it upon Aaron, and upon his garments, and upon his sons, and upon his sons' garments with him; and sanctified Aaron, and his garments, and his sons, and his sons' garments with him.

31 And Moses said unto Aaron and to his sons, Boil the flesh at the door of the tabernacle of the congregation: and there eat it with the bread that is in the basket of consecrations, as I commanded, saying, Aaron and his sons shall eat it.

32 And that which remaineth of the flesh and of the bread shall ye burn with fire.

33 And ye shall not go out of the door of the tabernacle of the congregation in seven days, until the days of your consecration be at an end: for seven days shall he consecrate you.

34 As he hath done this day, so the Lord hath commanded to do, to make an atonement for you.

35 Therefore shall ye abide at the door of the tabernacle of the congregation day and night seven days, and keep the charge of the Lord, that ye die not: for so I am commanded.

36 So Aaron and his sons did all things which the Lord commanded by the hand of Moses.

Exodus 27

And thou shalt make an altar of shittim wood, five cubits long, and five cubits broad; the altar shall be foursquare: and the height thereof shall be three cubits.

2 And thou shalt make the horns of it upon the four corners thereof: his horns shall be of the same: and thou shalt overlay it with brass.

3 And thou shalt make his pans to receive his ashes, and his shovels, and his basons, and

his fleshhooks, and his firepans: all the vessels thereof thou shalt make of brass.

4 And thou shalt make for it a grate of network of brass; and upon the net shalt thou make four brasen rings in the four corners thereof.

5 And thou shalt put it under the compass of the altar beneath, that the net may be even to the midst of the altar.

6 And thou shalt make staves for the altar, staves of shittim wood, and overlay them with brass.

7 And the staves shall be put into the rings, and the staves shall be upon the two sides of the altar, to bear it.

8 Hollow with boards shalt thou make it: as it was shewed thee in the mount, so shall they make it.

9 And thou shalt make the court of the tabernacle: for the south side southward there shall be hangings for the court of fine twined linen of an hundred cubits long for one side:

10 And the twenty pillars thereof and their twenty sockets shall be of brass; the hooks of the pillars and their fillets shall be of silver.

11 And likewise for the north side in length there shall be hangings of an hundred cubits long, and his twenty pillars and their twenty sockets of brass; the hooks of the pillars and their fillets of silver.

12 And for the breadth of the court on the west side shall be hangings of fifty cubits: their pillars ten, and their sockets ten.

13 And the breadth of the court on the east side eastward shall be fifty cubits.

14 The hangings of one side of the gate shall be fifteen cubits: their pillars three, and their sockets three.

15 And on the other side shall be hangings fifteen cubits: their pillars three, and their sockets three.

16 And for the gate of the court shall be an hanging of twenty cubits, of blue, and purple, and scarlet, and fine twined linen, wrought with needlework: and their pillars shall be four, and their sockets four.

17 All the pillars round about the court shall be filleted with silver; their hooks shall be of silver, and their sockets of brass.

18 The length of the court shall be an hundred cubits, and the breadth fifty every where, and the height five cubits of fine twined linen, and their sockets of brass.

19 All the vessels of the tabernacle in all the service thereof, and all the pins thereof, and all the pins of the court, shall be of brass.

20 And thou shalt command the children of Israel, that they bring thee pure oil olive beaten for the light, to cause the lamp to burn always.

21 In the tabernacle of the congregation without the vail, which is before the testimony, Aaron and his sons shall order it from evening to morning before the Lord: it shall be a statute for ever unto their generations on the behalf of the

children of Israel.

Exodus 37

And Bezaleel made the ark of shittim wood: two cubits and a half was the length of it, and a cubit and a half the breadth of it, and a cubit and a half the height of it:

2 And he overlaid it with pure gold within and without, and made a crown of gold to it round about.

3 And he cast for it four rings of gold, to be set by the four corners of it; even two rings upon the one side of it, and two rings upon the other side of it.

4 And he made staves of shittim wood, and overlaid them with gold.

5 And he put the staves into the rings by the sides of the ark, to bear the ark.

6 And he made the mercy seat of pure gold: two cubits and a half was the length thereof, and one cubit and a half the breadth thereof.

7 And he made two cherubims of gold, beaten out of one piece made he them, on the two ends of the mercy seat;

8 One cherub on the end on this side, and another cherub on the other end on that side: out of the mercy seat made he the cherubims on the two ends thereof.

9 And the cherubims spread out their wings on high, and covered with their wings over the mercy seat, with their faces one to another; even

to the mercy seatward were the faces of the cherubims.

10 And he made the table of shittim wood: two cubits was the length thereof, and a cubit the breadth thereof, and a cubit and a half the height thereof:

11 And he overlaid it with pure gold, and made thereunto a crown of gold round about.

12 Also he made thereunto a border of an handbreadth round about; and made a crown of gold for the border thereof round about.

13 And he cast for it four rings of gold, and put the rings upon the four corners that were in the four feet thereof.

14 Over against the border were the rings, the places for the staves to bear the table.

15 And he made the staves of shittim wood, and overlaid them with gold, to bear the table.

16 And he made the vessels which were upon the table, his dishes, and his spoons, and his bowls, and his covers to cover withal, of pure gold.

17 And he made the candlestick of pure gold: of beaten work made he the candlestick; his shaft, and his branch, his bowls, his knops, and his flowers, were of the same:

18 And six branches going out of the sides thereof; three branches of the candlestick out of the one side thereof, and three branches of the candlestick out of the other side thereof:

19 Three bowls made after the fashion of almonds in one branch, a knop and a flower;

and three bowls made like almonds in another branch, a knop and a flower: so throughout the six branches going out of the candlestick.

20 And in the candlestick were four bowls made like almonds, his knops, and his flowers:

21 And a knop under two branches of the same, and a knop under two branches of the same, and a knop under two branches of the same, according to the six branches going out of it.

22 Their knops and their branches were of the same: all of it was one beaten work of pure gold.

23 And he made his seven lamps, and his snuffers, and his snuffdishes, of pure gold.

24 Of a talent of pure gold made he it, and all the vessels thereof.

25 And he made the incense altar of shittim wood: the length of it was a cubit, and the breadth of it a cubit; it was foursquare; and two cubits was the height of it; the horns thereof were of the same.

26 And he overlaid it with pure gold, both the top of it, and the sides thereof round about, and the horns of it: also he made unto it a crown of gold round about.

27 And he made two rings of gold for it under the crown thereof, by the two corners of it, upon the two sides thereof, to be places for the staves to bear it withal.

28 And he made the staves of shittim wood, and overlaid them with gold.

29 And he made the holy anointing oil, and the

pure incense of sweet spices, according to the work of the apothecary.

Revelation 9

And the fifth angel sounded, and I saw a star fall from heaven unto the earth: and to him was given the key of the bottomless pit.

2 And he opened the bottomless pit; and there arose a smoke out of the pit, as the smoke of a great furnace; and the sun and the air were darkened by reason of the smoke of the pit.

3 And there came out of the smoke locusts upon the earth: and unto them was given power, as the scorpions of the earth have power.

4 And it was commanded them that they should not hurt the grass of the earth, neither any green thing, neither any tree; but only those men which have not the seal of God in their foreheads.

5 And to them it was given that they should not kill them, but that they should be tormented five months: and their torment was as the torment of a scorpion, when he striketh a man.

6 And in those days shall men seek death, and shall not find it; and shall desire to die, and death shall flee from them.

7 And the shapes of the locusts were like unto horses prepared unto battle; and on their heads were as it were crowns like gold, and their faces were as the faces of men.

8 And they had hair as the hair of women, and their teeth were as the teeth of lions.

9 And they had breastplates, as it were

breastplates of iron; and the sound of their wings was as the sound of chariots of many horses running to battle.

10 And they had tails like unto scorpions, and there were stings in their tails: and their power was to hurt men five months.

11 And they had a king over them, which is the angel of the bottomless pit, whose name in the Hebrew tongue is Abaddon, but in the Greek tongue hath his name Apollyon.

12 One woe is past; and, behold, there come two woes more hereafter.

13 And the sixth angel sounded, and I heard a voice from the four horns of the golden altar which is before God,

14 Saying to the sixth angel which had the trumpet, Loose the four angels which are bound in the great river Euphrates.

15 And the four angels were loosed, which were prepared for an hour, and a day, and a month, and a year, for to slay the third part of men.

16 And the number of the army of the horsemen were two hundred thousand thousand: and I heard the number of them.

17 And thus I saw the horses in the vision, and them that sat on them, having breastplates of fire, and of jacinth, and brimstone: and the heads of the horses were as the heads of lions; and out of their mouths issued fire and smoke and brimstone.

18 By these three was the third part of men

killed, by the fire, and by the smoke, and by the brimstone, which issued out of their mouths.

19 For their power is in their mouth, and in their tails: for their tails were like unto serpents, and had heads, and with them they do hurt.

20 And the rest of the men which were not killed by these plagues yet repented not of the works of their hands, that they should not worship devils, and idols of gold, and silver, and brass, and stone, and of wood: which neither can see, nor hear, nor walk:

21 Neither repented they of their murders, nor of their sorceries, nor of their fornication, nor of their thefts.

Hebrews 9

Then verily the first covenant had also ordinances of divine service, and a worldly sanctuary.

2 For there was a tabernacle made; the first, wherein was the candlestick, and the table, and the shewbread; which is called the sanctuary.

3 And after the second veil, the tabernacle which is called the Holiest of all;

4 Which had the golden censer, and the ark of the covenant overlaid round about with gold, wherein was the golden pot that had manna, and Aaron's rod that budded, and the tables of the covenant;

5 And over it the cherubims of glory shadowing the mercyseat; of which we cannot now speak particularly.

6 Now when these things were thus ordained, the priests went always into the first tabernacle, accomplishing the service of God.

7 But into the second went the high priest alone once every year, not without blood, which he offered for himself, and for the errors of the people:

8 The Holy Ghost this signifying, that the way into the holiest of all was not yet made manifest, while as the first tabernacle was yet standing:

9 Which was a figure for the time then present, in which were offered both gifts and sacrifices, that could not make him that did the service perfect, as pertaining to the conscience;

10 Which stood only in meats and drinks, and divers washings, and carnal ordinances, imposed on them until the time of reformation.

11 But Christ being come an high priest of good things to come, by a greater and more perfect tabernacle, not made with hands, that is to say, not of this building;

12 Neither by the blood of goats and calves, but by his own blood he entered in once into the holy place, having obtained eternal redemption for us.

13 For if the blood of bulls and of goats, and the ashes of an heifer sprinkling the unclean, sanctifieth to the purifying of the flesh:

14 How much more shall the blood of Christ, who through the eternal Spirit offered himself without spot to God, purge your conscience from

dead works to serve the living God?

15 And for this cause he is the mediator of the new testament, that by means of death, for the redemption of the transgressions that were under the first testament, they which are called might receive the promise of eternal inheritance.

16 For where a testament is, there must also of necessity be the death of the testator.

17 For a testament is of force after men are dead: otherwise it is of no strength at all while the testator liveth.

18 Whereupon neither the first testament was dedicated without blood.

19 For when Moses had spoken every precept to all the people according to the law, he took the blood of calves and of goats, with water, and scarlet wool, and hyssop, and sprinkled both the book, and all the people,

20 Saying, This is the blood of the testament which God hath enjoined unto you.

21 Moreover he sprinkled with blood both the tabernacle, and all the vessels of the ministry.

22 And almost all things are by the law purged with blood; and without shedding of blood is no remission.

23 It was therefore necessary that the patterns of things in the heavens should be purified with these; but the heavenly things themselves with better sacrifices than these.

24 For Christ is not entered into the holy places made with hands, which are the figures of the

true; but into heaven itself, now to appear in the presence of God for us:

25 Nor yet that he should offer himself often, as the high priest entereth into the holy place every year with blood of others;

26 For then must he often have suffered since the foundation of the world: but now once in the end of the world hath he appeared to put away sin by the sacrifice of himself.

27 And as it is appointed unto men once to die, but after this the judgment:

28 So Christ was once offered to bear the sins of many; and unto them that look for him shall he appear the second time without sin unto salvation.

Gold, Silver, Brass (Copper) Horns... Huh?

[Research on electricity frequency property of blood]

Maoqing Hu 1, Hua Huang, Zirun Yuan, Huaiqing Chen, Lihua Den

Abstract
On the basis of our previous work, the electric frequency property of human blood in different components, in physiological state and in pathological state (diabetes) are tested and analyzed in the range of 1Hz-20MHz progressively. Among the different components of blood; the lowest electrical impedance is

serum; the plasma and the whole blood gradually become larger, the blood corpuscle is the largest one. Otherwise, the negative phase of serum is the largest, the plasma and the whole blood are lower, and the blood corpuscle is the lowest. Here, the question is why the effect of the electric capacity of serum and plasma is the biggest in the condition of no cell and cell membrane; diabetes mellitus is an endocrine disorder in which blood changes obviously, the electric frequency property of the blood of diabetic patients changes markedly; the electrical impedance of blood decreases (more obviously with low frequency), the negative phase increases (more obviously with high frequency). These indicate that the increase of electric conductivity in diabetic patients' blood is due to electric capacitance conductivity that is related to the changes of cell membrane, deformation abilities and aggregation of RBC. Related experiments demonstrate again that with the progressing of research in the electric frequency property of blood, we may use the theory and method of electricity to examine some important characters of blood in a different way, and so to corroborate other tests and analyses.

Gold, Silver, Brass (Copper) Horns... Huh? Nah bruh, don't let these Master Degree GoofBalls nor the I read the Bible 6,000 times people get you!!! Read it for yourself! There are descriptions with Acacia Wood overlaid with Gold, Silver or Brass (Copper)!!! Do you understand the Periodic Table??? Silver, Copper & Gold are the **MOST ELECTRICALLY CONDUCTIVE METALS ON EARTH**!!! Yall think I chose to write all these books, do all these lectures, lose out on the loves of my life, time with my friends and family, then they stop talking to me or die... I died already, my heart stopped this was the deal I made for life and I am at peace with it... I only just had Toni at the time, this is Oct 28th 2001... I was 24, I am now 47, I get to see everyone in my age group that went down the roads I was on... Few are left, those that are left aren't OK. They seen to much violence and pain, they have been a part of too much violence and pain. I got to live, I got a second chance but I haven't quite escape the... My point is, this is serious work, my typos are a security system. Those reading these books for the wrong reasons, will tap out! This is really only for me and you, the people we choose to touch, more rather those that choose to touch us. You can't help but share this information with everyone you know but... They won't all receive it and that hurts the most, when you watch them kill themselves! Do you get it yet? Inspiration has two sources, two signals, Life or Death. Cold vs Heat, Positive Ions vs Negative Ions...

I had Lyrics like... I am so ashamed of my old

music most times... I can't even repeat them here but God brung me out, so I could share this truth with you today. I think the Blood on the Altar was a way of tuning the frequency, that way instead of the infinite sea of frequencies, the altars would be tuned to the Life Station. Yeah it's wild, I am not saying its cool or I approve, I am just stating what I know for you. There was no boogie man and God definitely wasn't the Devil. You know what's crazy, is people that want to attack the BuyBuild, use verses like these to turn people away. These are people that eat animals daily, talking to other people that eat animals daily. Like yeah oh God sacrifice a animal.... Devil.... Blah blah blah.... I be silently (or not so silently) thinking what does that make you? If you take dead animals cook them and eat them with your kids 2-4 times a day?

Acacia wood has some wild property to it too, it's rich in DMT. I am not suggesting you eat it, nor was the Bible. I am just telling you, wooden horns made from a Acacia dipped in Silver, Copper or Gold is clearly a Radio Wave Antenna! These Altars were built to communicate, get on his good side, commemorate moments of and to stay in tune with God. If you don't believe it, look the altars up for yourself:

Noah's Altar -

Abraham's Altars -

Isaac's Altar -

Jacob's Altars -

The Altars of Exodus -

Jehovah-Nissi -

An Altar of Earth -

An Altar of Stone -

Moses' Altar Beneath Sinai -

The Brazen Altar -

The Golden Altar -

Aaron's Altar -

Balaam's Altars -

The Altar of Ebal -

The Altar of Ed -

Gideon's Altars -

Manoah's Altar -

The Altar of Mizpeh -

David's Altar -

Elijah's Altar -

The Altars were all built on 'Hallowed' ground or a threshing floor! Here is the kicker, the threshing floor as you know, we discussed this 10,000 times, was built with 26 magnetic grains on it, 26! Twenty-six the number of God, 26 the number of Iron, 26 the number of Grains on the threshing floor and now we know 26 is the number of times the Bible has the term 'Horns of the Altar'. We also know that 26 is the only number that sits in between a square and a cube (reversed but remember the Bible language is read right to left). Five squared and three cubed, 25 & 27, that anomaly doesn't happen with any other number. A square and a cube represent 2 dimensions and 3 dimensions, 26 represents the bridge between the two, God. We are inside his image, as his image, which is binary, the electron and proton, male/female, electric/

magnetic etc...

Hallowed - Old English halgian "**to make holy**, sanctify; to honor as holy, consecrate, ordain," related to halig "**holy**," from Proto-Germanic *hailagon (source also of Old Saxon helagon, Middle Dutch heligen, Old Norse helga), from PIE root *kailo- "**whole**, uninjured, of good omen" (see health). Used in Christian translations to render Latin sanctificare.

"holy person, saint," Old English haliga, halga, from hallow (v.). Obsolete except in Halloween.

Hallowed Ground - Naturally magnetic, remember this is a big iron rock we are floating in space on! Iron, 26...

The other impossible anomaly (related to this) in the Bible is the speed of light. It's not just the magnetic foundations of these Altars with Horn Antennas or the word made flesh... It's the people are actually identified with Light as well...

Lets start here:

Jesus's event was on the 6th Day, 3 Days later on the 9th hour...

100% = 360 degrees = Completion

432 SQUARED = the approx speed of Light or 7 times the Globe a second.

432 = 9

186,624 = 9

Numbers 2:9 read backwards or right to left is 9 squared... LMAO... I'm reaching right? Watch this...

Numbers 2

And the Lord spake unto Moses and unto Aaron, saying,

2 Every man of the children of Israel shall pitch by his own standard, with the ensign of their father's house: far off about the tabernacle of the congregation shall they pitch.

3 And on the east side toward the rising of the sun shall they of the standard of the camp of Judah pitch throughout their armies: and Nahshon the son of Amminadab shall be captain of the children of Judah.

4 And his host, and those that were numbered of them, were threescore and fourteen thousand and six hundred.

5 And those that do pitch next unto him shall be the tribe of Issachar: and Nethaneel the son of Zuar shall be captain of the children of Issachar.

6 And his host, and those that were numbered thereof, were fifty and four thousand and four hundred.

7 Then the tribe of Zebulun: and Eliab the son of Helon shall be captain of the children of Zebulun.

8 And his host, and those that were numbered thereof, were fifty and seven thousand and four hundred.

9 **<u>All that were numbered in the camp of Judah were an hundred thousand and fourscore thousand and six thousand and four hundred, throughout their armies. These shall first set forth</u>**.

10 On the south side shall be the standard of the camp of Reuben according to their armies: and the captain of the children of Reuben shall be Elizur the son of Shedeur.

11 And his host, and those that were numbered thereof, were forty and six thousand and five hundred.

12 And those which pitch by him shall be the tribe of Simeon: and the captain of the children of Simeon shall be Shelumiel the son of

Zurishaddai.

13 And his host, and those that were numbered of them, were fifty and nine thousand and three hundred.

14 Then the tribe of Gad: and the captain of the sons of Gad shall be Eliasaph the son of Reuel.

15 And his host, and those that were numbered of them, were forty and five thousand and six hundred and fifty.

16 All that were numbered in the camp of Reuben were an hundred thousand and fifty and one thousand and four hundred and fifty, throughout their armies. And they shall set forth in the second rank.

17 Then the tabernacle of the congregation shall set forward with the camp of the Levites in the midst of the camp: as they encamp, so shall they set forward, every man in his place by their standards.

18 On the west side shall be the standard of the camp of Ephraim according to their armies: and the captain of the sons of Ephraim shall be Elishama the son of Ammihud.

19 And his host, and those that were numbered of them, were forty thousand and five hundred.

20 And by him shall be the tribe of Manasseh: and the captain of the children of Manasseh shall be Gamaliel the son of Pedahzur.

21 And his host, and those that were numbered of them, were thirty and two thousand and two hundred.

22 Then the tribe of Benjamin: and the captain of the sons of Benjamin shall be Abidan the son of Gideoni.

23 And his host, and those that were numbered of them, were thirty and five thousand and four hundred.

24 All that were numbered of the camp of Ephraim were an hundred thousand and eight thousand and an hundred, throughout their armies. And they shall go forward in the third rank.

25 The standard of the camp of Dan shall be on the north side by their armies: and the captain of the children of Dan shall be Ahiezer the son of Ammishaddai.

26 And his host, and those that were numbered of them, were threescore and two thousand and seven hundred.

27 And those that encamp by him shall be the tribe of Asher: and the captain of the children of

Asher shall be Pagiel the son of Ocran.

28 And his host, and those that were numbered of them, were forty and one thousand and five hundred.

29 Then the tribe of Naphtali: and the captain of the children of Naphtali shall be Ahira the son of Enan.

30 And his host, and those that were numbered of them, were fifty and three thousand and four hundred.

31 All they that were numbered in the camp of Dan were an hundred thousand and fifty and seven thousand and six hundred. They shall go hindmost with their standards.

32 These are those which were numbered of the children of Israel by the house of their fathers: all those that were numbered of the camps throughout their hosts were six hundred thousand and three thousand and five hundred and fifty.

33 But the Levites were not numbered among the children of Israel; as the Lord commanded Moses.

34 And the children of Israel did according to all that the Lord commanded Moses: so they pitched by their standards, and so they set forward,

every one after their families, according to the house of their fathers.

Yeah spooky!!! The Bible itself is spooky action at a distance, especially when you study it like this! We are entangled with each other and entangled with God!

Number 2:9 Is the speed of Light!!! What is that doing in there? Radio waves are the Largest Waves and they move at the speed of Light!!! The word is made flesh, flesh only the children of Israel have!!!

The Tribe of Light.

Numbers identify the Tribe with 186,400, which is the Constant for the Speed of Light. When the Bible was written, a couple thousand years ago, they couldn't have known...

Are you going to just explain away these things? Will you let some PHD guy tell you don't read it literally, cause 186,400 is really saying that TransActions are not sinful blah blah blah.... Aliens are really from other planets, they are here because their planets are in trouble blah blah blah... You know how many TV shows, movies and Trillions of Dollars went into this Brain Washing!!! If there are planets so far away they avoided our detection, and they have developed technology superior enough to get here, they would've preserved their planet!! Don't believe none of these alien stories! In May 1947, Parsons gave a talk at the Pacific Rocket Society in which he predicted that rockets would take humans to the Moon. Although he had become distant from the now largely defunct O.T.O. and had sold much of his Crowleyan library, he continued to correspond with Crowley until the latter's death in December 1947.

It must also be a coincidence that Crowley and Jack Parson's Father died in 1947? The same exact year that Alien Mania starts with Roswell and Area 51? Are you guys sleep at the wheel? Repeat after me...

Jesus take the wheel!!! Get a Rosary, wether you are baptized or not, it's wild out here! Do a whole rosary of prayers on Jesus taking the wheel! Yes of course I get the quadruple entendre, car wheel, vortex math, number of rosary beads, waveguide around the earth.... You need to tap into this ancient technology of yours! The devil will use it for you, if you don't.

How do you think we stay warm, 98 maybe 99 degrees? We constantly emit Infrared Light, all light can be converted into information. The planet constantly emits Infrared Light. The sun constantly emits Infrared Light.

Emit = Time? When you go completely cold, stop emission, your dead, time's up.

*Light, sound, time & space only exist in your mind.

*Traveling at Light Speed collapses Time & Space to Zero. Serotonin and Dopamine utilize this relationship to modulate your experience of time and space.

*Light is the bridge between energy and matter,

like 2 dimensions and 3 dimensions. Get it? The same way 26 sits between 2D and 3D, light sits between energy and matter.

*The universe is not limited by time or space but by Light, immaterial Light. Light which **measures** the same from any one **point** in space to any other! Light which **measures** the same from any one **point** in time to any other.

*The speed of light is how we interpret Light from stars (especially Radio waves).

*The Sun is 93 million miles away, so sunlight takes 8 and 1/3 minutes to get to us. This gives us 365 Day years and 7 day weeks, encoded in our blood as P.H. 7.365. It must also be a co-inky dink, that we have an almost 24 hour day, to match our almost 24 degree axis, huh? Mr. Parsons basically formed NASA when he was almost 24, I died at 24 (Ictal Asystole), I don't overlook these number things...

James 1:17

Every good thing given and every perfect gift is from above, coming down from the Father of lights, with whom there is no variation or shifting shadow.

Psalm 139:7-12

Where can I go from Your Spirit? Or where can

I flee from Your presence? If I ascend to heaven, You are there; If I make my bed in Sheol, behold, You are there. If I take the wings of the dawn, If I dwell in the remotest part of the sea, Even there Your hand will lead me, And Your right hand will lay hold of me. If I say, "Surely the darkness will overwhelm me, And the light around me will be night," Even the darkness is not dark to You, And the night is as bright as the day. Darkness and light are alike to You.

Remember there is no such thing as darkness, just light waves we don't have pigment or opsins for. Please read &/or reread the Speak it into Existence book and the Observer Effect book! I really feel like these books are the only way to really understand Melanin vs Diabetes book 1!!!

We have to ask ourselves... Is there a underlining truth to a book of words, that all have a numerical value, that document the creation of sound, being transformed into light, that light becoming space and time or fractions of the original piece with the Magnetic, housed in temples, **temples which have bleuprints we recognize as Genetics**? If you don't wake up, the Devil...

MIT News on Campus & around the World
Scientists control biological materials with

radio waves

Deborah Halber

It's not exactly "ET, phone home," but MIT researchers reported in the Jan. 10 issue of Nature that they can "speak" to DNA biomolecules with radio waves.

The goal is to instruct biological materials how to act for a variety of purposes. Biological machines may one day be used to perform computation, assemble computer components or become part of computer hardware or circuitry. Radio-controlled biology may lead to single-atom or single-molecule machines, or the ability to hook tiny antennae into living systems to turn genes on and off.

"Recent studies have provided new insights into the complexity, precision and efficiency of biomolecular machines at the molecular scale, inspiring the development of physical and chemical manipulation of biological systems," said Joseph M. Jacobson, associate professor at the Media Lab and one of the paper's authors. "Manipulation of DNA is interesting because it has been shown recently that is has potential as an actuator (a hard drive component) and can be used to perform computational operations."

The researchers predict that radio frequency (RF)

biology will have a broad range of applications. Because virtually all biological molecules can be linked with gold or other semiconducting nanoparticles, these molecules can be controlled electronically, remotely, reversibly and precisely, said Shuguang Zhang, associate director of the Center for Biomedical Engineering and another author of the study. Such systems will have profound implications for finely dissecting detailed molecular interactions and formations, he said.

SINGLE-ATOM MACHINES

Jacobson, head of the Media Lab's Molecular Machine group , has a background in quantum physics. He became interested in using biology as a tool to create nanometer-length machines. The ultimate goal, he said, is a single-atom or single-molecule machine.

It's hard to manufacture computer chips much smaller than 30 nanometers, but biology has an excellent track record at creating tiny workable systems. The cell itself is a phenomenal little machine with its own power supply and memory. "If we're interested in molecular-scale machines, biology is a wonderful place to start," Jacobson said.

He worked with researchers from the Center for Biomedical Engineering (CBE) to attach tiny radio-frequency antennae--a metal nanocluster of less than 100 atoms--to DNA. When a radio-frequency magnetic field is transmitted into the little antennae, the molecule is zapped with energy and responds.

Hybridization is the process of joining two complementary strands of DNA, or one each of DNA and RNA, to form a double-stranded molecule. In dehybridization, the strands unwind. Using this technique, the researchers dehybridized double-stranded DNA in a matter of seconds. The switching, which is reversible, did not affect neighboring molecules. - Go read the full articles if you want!!!

In fact pull out the Gold Book, and read our breakdown of Creation. There are no holes in the story of Genesis!

The Gold Book
L'Goat Book
The Horus & Set TransAction
The PDF book
The Osiris, Diabetes & Respiration book
The IB book
The Autism book
The Speak it into Existence book
The Observer Effect book
The Electrician's Radio Anatomy Book

Don't listen to those people that will tell you Faith needs to be blind, or you shouldn't study bible concepts, in books besides the Bible.

Matthew 7:7

7 Ask, and it shall be given you; seek, and ye shall find; knock, and it shall be opened unto you

Let me ask you something, what is the difference between the Observer Effect and how a Neuron works? I mean like... Watching makes waves of potential collapse into particles... In your body now, Billions of Neurons collapse waves into particles a zillion times a day! The wave is chemiluminescent and the particles are called Neuro**Transmitters**. Who observed the first wave to begin making particles, that began the initial formation of Matter?

QUICK QUESTIONS

If Radio, TV & the Internet run on Radio Waves, and a small group of people control the 'Media"...??? What are they actually in control of??? Baby development? I mean you been asking yourself for decades why is metal in 'jabs'? Better yet, let's just move forward, what exactly does aluminum do in the human brain? Is the a co-incidence between the aluminum accumulating in your brain, and neuromelanin loss? Alzheimers and Dementia are easily the biological version of 'bad signal'....

Aluminum and Alzheimer's disease: after a century of controversy, is there a plausible

link?

Lucija Tomljenovic 1

Abstract

The brain is a highly compartmentalized organ exceptionally susceptible to accumulation of metabolic errors. Alzheimer's disease (AD) is the most prevalent neurodegenerative disease of the elderly and is characterized by regional specificity of neural aberrations associated with higher cognitive functions. Aluminum (Al) is the most abundant neurotoxic metal on earth, widely bioavailable to humans and repeatedly shown to accumulate in AD-susceptible neuronal foci. In spite of this, the role of Al in AD has been heavily disputed based on the following claims: 1) bioavailable Al cannot enter the brain in sufficient amounts to cause damage, 2) excess Al is efficiently excreted from the body, and 3) Al accumulation in neurons is a consequence rather than a cause of neuronal loss. Research, however, reveals that: 1) very small amounts of Al are needed to produce neurotoxicity and this criterion is satisfied through dietary Al intake, 2) Al sequesters different transport mechanisms to actively traverse brain barriers, 3) incremental acquisition of small amounts of Al over a lifetime favors its selective accumulation in brain tissues, and 4) since 1911, experimental evidence has repeatedly demonstrated that chronic Al intoxication reproduces neuropathological hallmarks of AD. Misconceptions about Al bioavailability may have misled scientists regarding the significance of Al in the pathogenesis of AD. The hypothesis that Al significantly contributes to AD is built upon very solid experimental evidence and should not be dismissed. Immediate steps should be taken to lessen human exposure to Al, which may be the single most aggravating and avoidable factor related to AD.

Aluminum is widely used in the construction of antennas due to its advantageous properties that enhance performance, durability, and cost-effectiveness. This report explores the key reasons for the preference of aluminum in antenna manufacturing, focusing on its conductive properties, corrosion resistance, weight, and other significant factors.

Electrical Conductivity

One of the primary reasons aluminum is favored in antenna design is its excellent electrical conductivity. Aluminum provides a good balance of conductivity compared to other materials commonly used for antennas, such as copper. While copper is slightly more conductive, aluminum allows for a lighter structure at a lower cost, making it a practical choice for both commercial and amateur antennas. Its ability to effectively conduct radio waves ensures minimal signal loss, which is crucial for maintaining antenna efficiency and performance.

Lightweight and Structural Integrity

Aluminum is significantly lighter than many alternative metals, such as steel. This lightweight nature facilitates easier handling, installation, and overall deployment of antennas, especially in applications where mobile or portable antennas are necessary. The reduced weight also minimizes stress on mounting structures, providing better durability and longevity. Despite its lightness, aluminum exhibits substantial strength and can withstand various environmental conditions without compromising structural integrity.

Corrosion Resistance

Aluminum naturally forms a protective oxide layer when exposed to air, which significantly enhances its corrosion resistance. This quality is particularly beneficial for outdoor antennas that may be exposed to harsh weather conditions, including rain, humidity, and temperature fluctuations. The oxide layer protects the underlying metal, ensuring that aluminum remains functional and aesthetically pleasing for an extended period. Thus, aluminum antennas require less maintenance compared to those made from less corrosion-resistant materials.

Cost-Effectiveness

Compared to copper and other high-conductivity materials, aluminum is considerably less expensive, making it an attractive option for manufacturers and consumers alike. This cost savings can lead to more economical production of antennas, allowing for wider accessibility in both consumer and commercial markets. The combination of affordability with the necessary conductive and structural properties makes aluminum a practical choice for antenna design.

Performance in Various Applications

Aluminum antennas can be designed to perform effectively across a variety of frequency ranges, which can be particularly advantageous in applications such as wireless communication, satellite transmission, and amateur radio. The adaptability of aluminum in forming different shapes—along with its ability to be easily machined—allows engineers to design antennas tailored to specific use cases and performance requirements.

Conclusion

The use of aluminum in antennas is primarily driven by its favorable combination of conductivity, lightweight nature, corrosion resistance, and cost-effectiveness. These attributes contribute to the overall performance and durability of antennas, making aluminum an optimal choice for manufacturers across various sectors. As technology continues to advance, the applications for aluminum in antenna production are likely to expand, leveraging its beneficial properties to support innovative designs and improved communication capabilities.

Aluminum is used in 'jabs' primarily as an

adjuvant, a substance that enhances the body's immune response to the 'jab'. Its inclusion helps improve the effectiveness of 'jab', ensuring that they provide robust protection against diseases. Below is a detailed report on the role, mechanisms, and safety aspects of aluminum in 'jab'.

Role of Aluminum as an Adjuvant

Aluminum salts, such as aluminum hydroxide, aluminum phosphate, and aluminum potassium sulfate, have been utilized as adjuvants in 'jabs' for over 70 years. These compounds help create a stronger immune response by stimulating the immune system to recognize and respond to the pathogen component of the 'jab'. By utilizing aluminum, 'jabs 'can often achieve a higher level of immunity with fewer doses.

Mechanism of Action

When injected, aluminum salts promote the recruitment of immune cells such as macrophages and dendritic cells to the injection site. This influx of immune cells enhances the body's ability to identify and attack pathogens, allowing for a more vigorous and prolonged immune response. The presence of aluminum adjuvants ensures that 'jabs 'generate sufficient antibodies and memory cells, which are crucial for long-term immunity against specific infections.

Safety and Regulatory Oversight

Extensive research and monitoring have been conducted regarding the safety of aluminum in 'jabs'. Regulatory agencies such as the CDC and FDA continuously assess and monitor 'jab' safety after clinical trials. Studies indicate that the small amounts of aluminum found in 'jabs '(typically ranging from 0.125 mg to 0.5 mg per dose) are well tolerated by the body and are much lower than the aluminum exposure that occurs naturally from food and environmental sources.

Natural Exposure: Infants typically ingest more aluminum through diet than they receive from 'jabs'; breastfed infants may ingest about 7 mg of aluminum within the first six months of life, while aluminum content in formula can be significantly higher. Thus, the cumulative aluminum burden from 'jabs 'remains within safe limits as established by scientific assessments.

Public Health Perspective

The persistent use of aluminum as a 'jab' adjuvant has contributed significantly to the effectiveness of 'jab 'programs worldwide. It enables 'jabs 'to elicit a strong immune response while minimizing the number of doses required for effective immunization. This not only reduces healthcare costs but also enhances compliance with 'jab 'schedules, ultimately leading to better public health outcomes.

Recent Research and Developments

While aluminum adjuvants are considered safe, some recent studies have suggested potential associations between aluminum exposure from 'jabs 'and certain health outcomes, such as asthma. However, these findings have not led to changes in 'jab 'recommendations, as further investigation is deemed necessary, and the correlation is not yet established as causative.

Conclusion

Aluminum plays a critical role in enhancing the efficacy of vaccines as an adjuvant. Its use is supported by decades of research demonstrating safety and effectiveness in promoting strong immune responses. Ongoing surveillance and research continue to ensure that the benefits of 'jabs 'that include aluminum far outweigh any potential risks, reinforcing their importance in public health initiatives.

Is it possible that the Fluoride and Aluminum work together, to make you more susceptible to Radio Waves? I mean we convert infrared into melatonin, so how does Fluoride block infrared?

Fluoride- and electromagnetic radiation-induced genotoxicity and impaired melatonin secretion

Bruce Spittle

SUMMARY: Rao and Thakur have shown that the antioxidants melatonin and alma
(Emblica officinales, Indian gooseberry) are effective, both individually and in
combination, against fluoride-induced genotoxicity in human peripheral blood

lymphocyte cells, which was first described in humans in 1994. Some animal and human work also suggests that fluoride (F) can impair the defensive response to genotoxicity by being deposited in high concentrations in the pineal gland and, through an enzyme-inhibiting action, reducing the secretion of melatonin, a powerful antioxidant able to eliminate free radicals and protect DNA. In having the capacity to be both genotoxic and impair melatonin secretion, F is similar to electromagnetic radiation, at power line frequencies and above, and both have very low or zero thresholds for causing toxicity. In view of the seriousness of neoplasia, the effect of fluoride on melatonin secretion warrants further research

STARS

Radio Stars: imaging stellar surfaces at radio wavelengths

Stars emit radio waves through a wide variety of physical processes, depending on the type of star, its age, and its physical properties (e.g., mass, temperature, and chemical composition). Observations of these various types of stellar radio waves provide astronomers with unique insights into how stars are born, evolve, interact with their environments, and eventually die.

Apart from the Sun, most other stars appear from Earth as mere pinpoints of light. However, using powerful arrays of radio interferometers such as the **Karl G. Jansky Very Large Array** in New Mexico and the **Atacama Millimeter/submillimeter Array (ALMA)** in northern Chile, it becomes possible to resolve the surfaces of some of the biggest and nearest stars known as red giants.

The **radio telescope** comprises 27 independent antennas in use at a given time plus one spare, each of which has a dish diameter of 25 meters (82 feet) and weighs 209 metric tons (230 short tons). The antennas are distributed along the three arms of a track, shaped in a wye (or Y) - configuration, (each of which measures 21 kilometres (13 mi) long). Using the rail tracks that follow each of these arms—and that, at one point, intersect with U.S. Route 60 at a level crossing—and a specially designed lifting locomotive ("Hein's Trein"), the antennas can be physically relocated to a number of prepared positions, allowing aperture synthesis interferometry with up to 351 independent baselines: in essence, the array acts as a single antenna with a variable diameter. The **angular resolution** that can be reached is between 0.2 and 0.04 arcseconds.

The world's largest telecommunications system. The Deep Space Network—or DSN—is NASA's international array of giant radio antennas that supports interplanetary spacecraft missions, plus a few that orbit Earth.

The NASA **Deep Space Network** (DSN) is a worldwide network of spacecraft communication ground segment facilities, located in the United States (California), Spain (Madrid), and Australia (Canberra), that supports NASA's interplanetary spacecraft missions. It also performs radio and radar astronomy observations for the exploration of the Solar System and the universe, and supports selected Earth-orbiting missions. **DSN is part of the NASA Jet Propulsion Laboratory (JPL)**.

Do you see what the satanic rituals was about???? Radio!!! **JPL was developing what**? They aint traveling into outer space they are traveling into inner space, your soul! There is no secret material only melanin polymers!!! There are no aliens, only demons!!! There is no outer space only inner space!!!

When we began to discuss Anansi the spider as electromagnetism, you may have overlooked that the web is inside you!

Arachnoid - "cobweb-like," especially of the membrane around the brain and spinal cord, 1789, from Modern Latin arachnoides, from Greek arakhnoeides "cobweb-like," from arakhnē "cobweb" (see arachnid) + -oeidēs (see -oid).

Arachnoid Mater - is one of the three meninges, the protective membranes that cover the brain and spinal cord. It is so named because of its resemblance to a spider web. The arachnoid mater is a derivative of the neural crest mesoectoderm in the embryo. The arachnoid mater is named after the Greek word arachne ("spider"), the suffix -oid ("in the image of"), and the Latin word mater ("mother"), because of the fine spider-web–like appearance of the delicate fibres of the arachnoid (arachnoid trabeculae) which extend down through the subarachnoid space and attach to the pia mater.

CSF circulates in the subarachnoid space (between arachnoid and pia mater). Cerebrospinal fluid is produced by the choroid plexus (inside the ventricles of the brain, which are in direct communication with the subarachnoid space so the CSF can flow freely through the nervous system). Cerebrospinal fluid is a transparent, colourless fluid and it is produced at about 500 ml/day. Its electrolyte levels, glucose levels, and pH are very similar to those in plasma, but the presence of blood in cerebrospinal fluid is always abnormal.

Don't forget this whole thing started with the Heart! the heart has 40,000 neurons and the ability to process, learn, and remember. The heart has an electrical component about 60 times greater than the brain, more importantly that electrical activity is the genesis of an electromagnetic energy field, 5000 times greater than the brain's, how much of that is Radio Waves? Did you know that studies in people suggest that, radio frequency radiation can raise the risk for cardiovascular disease by increasing blood pressure, total cholesterol and low-density lipoprotein cholesterol. Hello... Taps mic... Is this thing on? Does Black mean God's Children? Are Black diseases caused by being ignorant of this fact (if it's a fact)? I mean studies show that Radio Waves causes your body to act 'Black', by increasing blood pressure, total cholesterol and low-density lipoprotein cholesterol.

Gold Silver & Copper promote health... Aluminum...not so much, is this due to the signals? Have you seen the Omen? Damien was already born in real life. The movie the Omen was released in 1976, many people thought they had a opportunity to stop the antichrist lmao.... The AntiChrist wasn't Crowley, it was Jack Parsons. Jack Parsons was born to finish the Great Work of Thelema. The word θέλημα (thelema) is rare in Classical Greek, where it "signifies the appetitive will: desire, sometimes even sexual", but it is frequent in the Septuagint. Early Christian writings occasionally use the word to refer to the human will, and even the will of the Devil, but it usually refers to the will of God. In the Renaissance, a character named "Thelemia" represents will or desire in the Hypnerotomachia Poliphili of the Dominican friar and writer Francesco Colonna. The protagonist Poliphilo has two allegorical guides, Logistica (reason) and Thelemia (will or desire). When forced to choose, he chooses fulfillment of his sexual will over logic.

Mass Broadcasting would never be the same! Do your homework and go find out why all radio is Government? It's easy to use lyrics against you, they are federal since 1916 ish...

Listen I died at 24 years old, you may be saying "WE KNOW ALREADY ENQI"... The reason why I am being so repetitive is simple, we have many people that have flatlined and came back! Why is it hard to believe someone did it 2000 years ago? If God exists why is there war, starvation, disease etc... Easy because the Devil exists and HAS A BETTER TEAM!!! People that expect a lot from God do so while smoking poison, popping pills, eating garbage, they never pray, the never meditate, if they do it's for devilish desires! Satanist work hard and diligently, **THEY KNOW WHAT THEY ARE UP AGAINST AND WORK TOGETHER**!

We are supposed to be teaching the world, that includes Gentiles & Muslims!

Sahih 2767

When it will be the Day of Resurrection Allah would deliver to every Muslim a Jew or a Christian and say: That is your rescue from Hell-Fire.

There would come people amongst the Muslims on the Day of Resurrection with as heavy sins as a mountain, and Allah would forgive them and He would place in their stead the Jews and the Christians. (As far as I think), Abu Raub said: I do not know as to who is in doubt. Abu Burda said: I narrated it to 'Umar b. 'Abd al-'Aziz, whereupon he said: Was it your father who narrated it to you from Allah's Apostle (ﷺ)? I said: Yes.

Do understand it was the Muslim who started the TransAtlantic Slave Trade by enslaving, raping & killing off the Black Egyptians!

Jack Parsons the Antichrist was born Oct 2nd 1914 but the sins of the Flesh officially began long before then!!! King James was a rare breed because refused Corpse Medicine, that's why I support him and his Bible. Charles the 2nd, Francis the 1st etc… were all promoters of corpse medicine! If you ask me, this was the precursor to today's Human Trafficking and Abortion Clinic Menus! People voting for Kamala voted to continue this practice. You thought they were just throwing the babies in the trash? OOoohhh wait you thought they was burying the aborted babies in the grave yard? Better yet, you simply don't

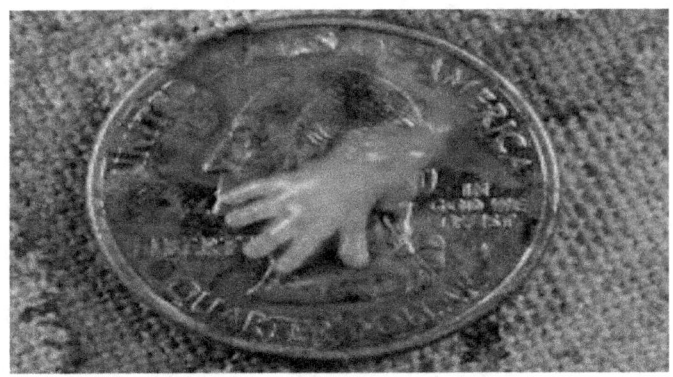

care, women hop up on the table and don't care whats done with their babies! Well they are sold, even the tiniest ones!

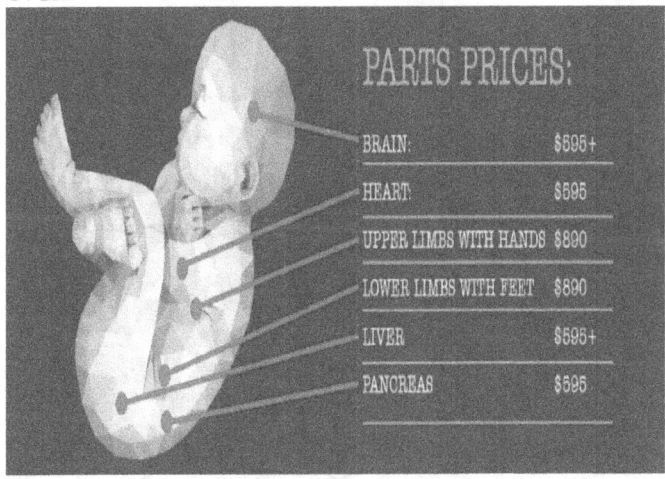

Fighting for your right to Humane Human Trafficking huh? Oh yeah, this is Satanism at it's finest and who is buying all these babies? Let me tell you this before you even say scientific researchers, they only need cells once. Once they

have cells they clone them and have enough for 100 lifetimes of tests. The babies are going to someone or something that constantly needs new babies to consume. Black People are working for Satan and the biggest human Traffickers on Earth!
There is about 700,000 abortions and 700,000 missing kids reported each year, the missing children can be found. The aborted children, never had a chance. Don't think this is Race exclusive either, the true Father of Pansexualist Satanism isn't Crowley it's a Blackman named Paschal Beverly Randolph. Helen Blavatsky was a student of his, he a student of Pansexualism and the Chakras. In September 1855, after "a long tour in Europe and Africa," he gave a public lecture to African Americans in NY on the subject of **emigrating to India**. Randolph believed that "the Negro is destined to extinction" in the United States.

Dr. York, Bobby Hemmit and many others studied Beverly to develop their Pansexual philosophies. Beverly was a believer in pre-Adamism (the belief that humans existed on earth before the biblical Adam) and wrote the book Pre-Adamite Man: demonstrating the existence of the human race upon the earth 100,000 thousand years ago! These were the Pansexuals mind you, they are in the Bible too. When Cain went to the land of Nod... them sodomites etc... The manner in which Randolph incorporated sex into his occult system was considered uncharacteristically bold for the period in which he lived. He believed that sex magic could lead to increased health, love, the empowerment of women, and children of superior intelligence. In his more underground publications, he wrote that church and marriage were oppressive forces that could be overthrown with the power of love in a world-wide revolution.

Randolph held an unusually expansive view of gender identity, considering earthly gender to be "provisional," and referring to God as both male and female. In a book on love he wrote:
I believe in love, all the way through. And while I live will help every man, woman, and the betweenities to win, obtain, intensify, deepen, purify, strengthen and keep it, and I will help all others to do the same. There! That's me! I mean it!

Galen & Hippocrates promoted Corpse Medicine, the eating of People (Black Bodies were the most valuable), strictly for health purposes though. 820 CE Arab Governor of Cairo Breaks Into the Great Pyramid known as the Caliph al Ma'mun. This wasn't enough, they enslaved the black Egyptians and ate the mummies! Thousands of years of history in the bellies of the Arabic-Moorish-Muslims!!! Rhazes, the Westernized name of Abu Bakr Muhammad ibn Zakariya al-Razi, a Persian scientist, physician, and Platonist philosopher in the Islamic age of enlightenment. He fostered the expansion of empirical medical knowledge and practical procedures for treatment, rather than theoretical reflections on illness and health. His contributions to public health included a recommended behavioral regimen to preserve good health and, in the field of clinical communicable disease control, criteria to distinguish smallpox from measles and much else. He changed the definition of mummy from mineral pitch to the wrapped corpses we know of today. The Arabs and Persians were the first to eat Black Bodies as 'medicine'. This is the medicine they introduced to Europe! Europe loved it, this was the 3rd largest income for the Muslim culture. The first was stolen Egyptian Relics, the second was enslaved Egyptians!The Medieval English term "mummy" was defined as "medical preparation of the substance of mummies", rather than the entire corpse, with Richard Hakluyt in 1599 AD complaining that "these dead bodies are the Mummy which the Phisistians and Apothecaries doe against our wills **make us to swallow**". These substances were called mummia.

In India today they still see Black Skin as less than human, this is so they are comfortable with you as a food source. The Lion will never see the

Deer as a equal, even if trained not to eat them.

This is the school of Alchemia these guys come from, they studied 'Blackness' all right... They wanted that flesh so bbad they could taste it, literally! It don't stop there, gunpowder, also commonly known as **Black Powder** to distinguish it from modern smokeless powder, is the earliest known chemical explosive. It consists of a mixture of sulfur, charcoal (which is mostly carbon), and potassium nitrate (saltpeter).

As a kid the antichrist was deeply into the Arabian Nights books, at 12 years old Jack Parsons successfully invoked the Devil in his bedroom, it is around this time he starts blowing thing up with GunPowder. It wasn't until a decade later he learned about liquid explosives from Eugen Sanger. The entire time he was a dedicated Satanist. He eventually founded two companies: the Jet Propulsion Laboratory (JPL) and the Aerojet Engineering Corporation. The JPL would go on to develop Radio Wave technology and the Aerojet company would go on to develop Infrared Technology. Both companies worked exclusively for the military! Do you get it? Television (TV) is a telecommunication medium for transmitting moving images and sound. Additionally, the term can refer to a physical television set rather than the medium of transmission. Television is a mass medium for advertising, entertainment, news, and sports. The medium is capable of more than "radio broadcasting," which refers to an audio signal sent to radio receivers.

Television became available in crude experimental forms in the 1920s, but only after several years of further development was the new technology marketed to consumers. After World War II, an improved form of black-and-white television broadcasting became popular in the United Kingdom and the United States, and television sets became commonplace in homes, businesses, and institutions. During the 1950s, television was the primary medium for influencing public opinion.

Everything changed post Jack... post Crowley... Jack was arguably a Billionaire, the character Tony Starks is loosely based on. A Star...

God made the Stars for What? I mean the word is in the Bible 68 times... There seems to be no specific point where the Bible focuses on the Gases or physical star, just their Light. That's the first thing to get, before you trade become stellar to be a Star on Earth.

Job 38

Then the Lord answered Job out of the whirlwind, and said,

2 Who is this that darkeneth counsel by words without knowledge?

3 Gird up now thy loins like a man; for I will demand of thee, and answer thou me.

4 Where wast thou when I laid the foundations of the earth? declare, if thou hast understanding.

5 Who hath laid the measures thereof, if thou knowest? or who hath stretched the line upon it?

6 Whereupon are the foundations thereof fastened? or who laid the corner stone thereof;

7 **When the morning stars sang together, and all the sons of God shouted for joy**?

8 Or who shut up the sea with doors, when it brake forth, as if it had issued out of the womb?

9 When I made the cloud the garment thereof, and thick darkness a swaddlingband for it,

10 And brake up for it my decreed place, and set bars and doors,

11 And said, Hitherto shalt thou come, but no further: and here shall thy proud waves be stayed?

12 Hast thou commanded the morning since thy days; and caused the dayspring to know his place;

13 That it might take hold of the ends of the earth, that the wicked might be shaken out of it?

14 It is turned as clay to the seal; and they stand as a garment.

15 And from the wicked their light is withholden, and the high arm shall be broken.

16 Hast thou entered into the springs of the sea? or hast thou walked in the search of the depth?

17 Have the gates of death been opened unto thee? or hast thou seen the doors of the shadow of death?

18 Hast thou perceived the breadth of the earth? declare if thou knowest it all.

19 Where is the way where light dwelleth? and as for darkness, where is the place thereof,

20 That thou shouldest take it to the bound thereof, and that thou shouldest know the paths to the house thereof?

21 Knowest thou it, because thou wast then born? or because the number of thy days is great?

22 Hast thou entered into the treasures of the snow? or hast thou seen the treasures of the hail,

23 Which I have reserved against the time of trouble, against the day of battle and war?

24 By what way is the light parted, which scattereth the east wind upon the earth?

25 Who hath divided a watercourse for the overflowing of waters, or a way for the lightning of thunder;

26 To cause it to rain on the earth, where no man is; on the wilderness, wherein there is no man;

27 To satisfy the desolate and waste ground; and to cause the bud of the tender herb to spring forth?

28 Hath the rain a father? or who hath begotten the drops of dew?

29 Out of whose womb came the ice? and the hoary frost of heaven, who hath gendered it?

30 The waters are hid as with a stone, and the face of the deep is frozen.

31 Canst thou bind the sweet influences of Pleiades, or loose the bands of Orion?

32 Canst thou bring forth Mazzaroth in his season? or canst thou guide Arcturus with his sons?

33 Knowest thou the ordinances of heaven? canst thou set the dominion thereof in the earth?

34 Canst thou lift up thy voice to the clouds, that

abundance of waters may cover thee?

35 Canst thou send lightnings, that they may go and say unto thee, Here we are?

36 Who hath put wisdom in the inward parts? or who hath given understanding to the heart?

37 Who can number the clouds in wisdom? or who can stay the bottles of heaven,

38 When the dust groweth into hardness, and the clods cleave fast together?

39 Wilt thou hunt the prey for the lion? or fill the appetite of the young lions,

40 When they couch in their dens, and abide in the covert to lie in wait?

41 Who provideth for the raven his food? when his young ones cry unto God, they wander for lack of meat.

What does this mean: **When the morning stars sang together, and all the sons of God shouted for joy**.

Asteroseismology - the study of oscillations in stars. Stars have many resonant modes and frequencies, and the path of sound waves passing through a star depends on the local speed of sound, which in turn depends on local temperature and chemical composition. Because the resulting oscillation modes are sensitive to different parts of the star, they inform astronomers about the internal structure of the star, which is otherwise not directly possible from overall properties like brightness and surface temperature.

1 Corinthians 15

Moreover, brethren, I declare unto you the gospel which I preached unto you, which also ye have received, and wherein ye stand;

2 By which also ye are saved, if ye keep in memory what I preached unto you, unless ye have believed in vain.

3 For I delivered unto you first of all that which I also received, how that Christ died for our sins according to the scriptures;

4 And that he was buried, and that he rose again the third day according to the scriptures:

5 And that he was seen of Cephas, then of the twelve:

6 After that, he was seen of above five hundred brethren at once; of whom the greater part remain unto this present, but some are fallen asleep.

7 After that, he was seen of James; then of all the

apostles.

8 And last of all he was seen of me also, as of one born out of due time.

9 For I am the least of the apostles, that am not meet to be called an apostle, because I persecuted the church of God.

10 But by the grace of God I am what I am: and his grace which was bestowed upon me was not in vain; but I laboured more abundantly than they all: yet not I, but the grace of God which was with me.

11 Therefore whether it were I or they, so we preach, and so ye believed.

12 Now if Christ be preached that he rose from the dead, how say some among you that there is no resurrection of the dead?

13 But if there be no resurrection of the dead, then is Christ not risen:

14 And if Christ be not risen, then is our preaching vain, and your faith is also vain.

15 Yea, and we are found false witnesses of God; because we have testified of God that he raised up Christ: whom he raised not up, if so be that the dead rise not.

16 For if the dead rise not, then is not Christ

raised:

17 And if Christ be not raised, your faith is vain; ye are yet in your sins.

18 Then they also which are fallen asleep in Christ are perished.

19 If in this life only we have hope in Christ, we are of all men most miserable.

20 But now is Christ risen from the dead, and become the firstfruits of them that slept.

21 For since by man came death, by man came also the resurrection of the dead.

22 For as in Adam all die, even so in Christ shall all be made alive.

23 But every man in his own order: Christ the firstfruits; afterward they that are Christ's at his coming.

24 Then cometh the end, when he shall have delivered up the kingdom to God, even the Father; when he shall have put down all rule and all authority and power.

25 For he must reign, till he hath put all enemies under his feet.

26 The last enemy that shall be destroyed is death.

27 For he hath put all things under his feet. But when he saith all things are put under him, it is manifest that he is excepted, which did put all things under him.

28 And when all things shall be subdued unto him, then shall the Son also himself be subject unto him that put all things under him, that God may be all in all.

29 Else what shall they do which are baptized for the dead, if the dead rise not at all? why are they then baptized for the dead?

30 And why stand we in jeopardy every hour?

31 I protest by your rejoicing which I have in Christ Jesus our Lord, I die daily.

32 If after the manner of men I have fought with beasts at Ephesus, what advantageth it me, if the dead rise not? let us eat and drink; for to morrow we die.

33 Be not deceived: evil communications corrupt good manners.

34 Awake to righteousness, and sin not; for some have not the knowledge of God: I speak this to your shame.

35 But some man will say, How are the dead raised up? and with what body do they come?

36 Thou fool, that which thou sowest is not quickened, except it die:

37 And that which thou sowest, thou sowest not that body that shall be, but bare grain, it may chance of wheat, or of some other grain:

38 But God giveth it a body as it hath pleased him, and to every seed his own body.

39 All flesh is not the same flesh: but there is one kind of flesh of men, another flesh of beasts, another of fishes, and another of birds.

40 There are also celestial bodies, and bodies terrestrial: but the glory of the celestial is one, and the glory of the terrestrial is another.

41 **There is one glory of the sun, and another glory of the moon, and another glory of the stars: for one star differeth from another star in glory**.

42 **So also is the resurrection of the dead. It is sown in corruption; it is raised in incorruption**:

43 **It is sown in dishonour; it is raised in glory: it is sown in weakness; it is raised in power**:

44 **It is sown a natural body; it is raised a spiritual body. There is a natural body, and there is a spiritual body**.

45 **And so it is written, The first man Adam was**

made a living soul; the last Adam was made a quickening spirit.

46 **Howbeit that was not first which is spiritual, but that which is natural; and afterward that which is spiritual.**

47 **The first man is of the earth, earthy; the second man is the Lord from heaven.**

48 **As is the earthy, such are they also that are earthy: and as is the heavenly, such are they also that are heavenly.**

49 **And as we have borne the image of the earthy, we shall also bear the image of the heavenly.**

50 **Now this I say, brethren, that flesh and blood cannot inherit the kingdom of God; neither doth corruption inherit incorruption.**

51 Behold, I shew you a mystery; We shall not all sleep, but we shall all be changed,

52 In a moment, in the twinkling of an eye, **at the last trump**: for the trumpet shall sound, **and the dead shall be raised incorruptible**, and we shall be changed.

53 For this corruptible must put on incorruption, and this mortal must put on immortality.

54 So when this corruptible shall have put on incorruption, and this mortal shall have put on immortality, then shall be brought to pass the saying that is written, Death is swallowed up in victory.

55 O death, where is thy sting? O grave, where is thy victory?

56 The sting of death is sin; and the strength of sin is the law.

57 But thanks be to God, which giveth us the victory through our Lord Jesus Christ.

58 Therefore, my beloved brethren, be ye stedfast, unmoveable, always abounding in the work of the Lord, forasmuch as ye know that your labour is not in vain in the Lord.

When your Heart is weighed… will you be raised?

IRON

I want you to put on your thinking Cap for this one! Remember the core of the planet is Iron and so is **YOURS**!

Iron is used in various aspects of radio transmission, primarily in the construction of antennas and magnetic components for communication equipment. Iron's properties make it beneficial for improving signal quality and enhancing the performance of radio frequency (RF) systems. Below is a detailed report on the uses of iron in radio transmission, addressing its applications, advantages, and associated considerations.

Applications of Iron in Radio Transmission

Antenna Construction:

Wire and Cores: Iron is often utilized in the production of wire antennas, where it can enhance the strength and durability of the antenna structure. Additionally, iron cores are employed in inductors and transformers used within radios to improve magnetic coupling and efficiency.

Ferrite Materials:

Ferrite, a ceramic compound of iron oxide combined with other metals, is widely used in RF applications. Ferrite cores can be found in antennas, transformers, and inductors, serving to concentrate magnetic fields and reduce loss at high frequencies. This is particularly crucial in applications like impedance matching and signal amplification.

Magnetic Shields:

Iron is also used to shield and isolate sensitive electronic components in radio systems. Magnetic shielding helps prevent interference from external RF signals and improves the clarity of transmitted signals.

Advantages of Using Iron

Magnetic Properties:

Iron exhibits significant magnetic permeability, making it effective in tasks that require the concentration of magnetic fields, which is important for efficient signal transmission and reception. This characteristic helps improve the performance of inductors, transformers, and other RF components.

Cost-Effectiveness:

Compared to other metals like copper, iron is relatively inexpensive, allowing for cost-effective production of antennas and associated equipment.

Durability and Strength:

Structures made from iron can withstand environmental stressors like wind and moisture, making iron a suitable choice for outdoor antennas.

Considerations and Limitations

While iron offers numerous benefits in radio transmission, some limitations must also be acknowledged:

Corrosion:

Iron is susceptible to rust and corrosion, particularly when exposed to moisture and outdoor environments. This necessitates protective coatings or treatments to increase longevity.

Conductivity:

While iron is conductive, it is less so compared to copper and aluminum. In applications where conductivity is paramount (like in transmitting antennas), aluminum or copper may be favored to minimize losses.

Weight:

Iron is heavier than other materials commonly used in antenna design, which may impact the installation and structural requirements for certain applications.

Conclusion

Iron plays a significant role in various components of radio transmission systems, from antennas to magnetic components. Its unique property profile—combining good magnetic characteristics, durability, and cost-effectiveness—makes it a valuable material in the field of radio communications. However, considerations regarding corrosion and weight often lead engineers to balance iron's benefits with materials like aluminum and copper in specific applications. Understanding these characteristics allows for better design and functionality in radio transmission technologies.

Ever wonder what's beneath Radio Waves, Low Frequency Electromagnetic Fields. Extremely low frequency (ELF) electric and magnetic fields (EMF) occupy the lower part of the electromagnetic spectrum in the frequency range 0-100 kHz. ELF EMF result from electrically charged particles. Even though

THESE ARE TECHNICALLY STILL CONSIDERED RADIO WAVES!
ElectronVolts - an electronvolt (symbol eV, also written electron-volt and electron volt) is the measure of an amount of **kinetic energy** gained by a single electron accelerating from rest through an electric potential difference of one volt in vacuum, 1 eV equal to the exact value $1.602176634 \times 10^{-19}$ J, a unit of energy or work, **the work required to move an electron** through a potential difference of one volt. 1 eV would correspond to an infrared photon of wavelength 1240 nm or frequency 241.8 THz.

4-12 ELECTRON VOLTS = UVC 100-320 NM DEATH WM

3.9 ELECTRON VOLTS = UVB SKIN 290-320 NM DISEASE JW

3.4 ELECTRON VOLTS = UVA 320-400 NM EYE DISEASE SW

.001 ELECTRON VOLTS = FAR INFRARED 1000000 NM BEYOND THIS POINT IS RADIO WAVES

.4 ELECTRON VOLTS = NEAR FAR INFRARED 3000 NM

.8 ELECTRON VOLTS = MEDIUM INFRARED 1500 NM

1.5 ELECTRON VOLTS = NEAR INFRARED 780

NM

1.7 ELECTRON VOLTS = VISIBLE RED 620-780 NM

2 ELECTRON VOLTS = VISIBLE ORANGE 585-620 NM

2.1 ELECTRON VOLTS = VISIBLE YELLOW 570-585 NM

2.3 ELECTRON VOLTS = VISIBLE GREEN 490-570 NM

2.6 ELECTRON VOLTS = VISIBLE BLUE 440-490 NM

2.9 ELECTRON VOLTS = VISIBLE INDIGO 420-440 NM

3 ELECTRON VOLTS = VISIBLE VIOLET 400-420 NM

We will look at the E.M. Spectrum in terms of Electron Volts, this makes more sense because all of chemistry is based on the movement of electrons. The Sun is the great visible agent of the first cause. This of course means you have to rethink the whole entire ElectroChemistry aka Dr. Sebi vs Dr. EnQi book...

Sound Range: 20 to 20,000 Hz

Voice Range 90 to 255 Hz

Radio Range: 1 hertz up to 3,000 billion

hertz. Below Radio is Cellular, Extremely low frequency (ELF) electric and magnetic fields (EMF) occupy the lower part of the electromagnetic spectrum in the frequency range 0-100 kHz. ELF EMF result from electrically charged particles.

Infrared Range 1 Trillion Hertz to ... this is where our body heat is...
I have to be repetitive so you get it! This would be the perfect time to pull out your Photochemistry and Black Lines Matters books! Every living cell contains electrically charged components, including ions (such as sodium, potassium, calcium, and chloride). As these ions move across cell membranes, they create electrical potentials and currents, leading to electrogenic processes such as action potentials in nerve and muscle cells. Various metabolic activities, including those involved in cellular respiration and muscle contractions, also produce electrical activity. The bioelectric signals emerging from these processes contribute to the overall electromagnetic landscape of the body. The Human Electric Activity, creates Low-frequency electromagnetic fields, which fall towards the lower end of the electromagnetic spectrum, typically ranging from 0 Hz up to around 30 kHz. Starting at 0, that means what? The Spark of Life I showed you in the Let there be Music book is **REALLY THE BEGINNING OF LIFE**!

PSALMS 82

God standeth in the congregation of the mighty; he judgeth among the gods.
2 How long will ye judge unjustly, and accept the persons of the wicked? Selah.
3 Defend the poor and fatherless: do justice to the afflicted and needy.
4 Deliver the poor and needy: rid them out of the hand of the wicked.
5 They know not, neither will they understand; they walk on in darkness: all the foundations of the earth are out of course.
6 **I have said, Ye are gods; and all of you are children of the most High**.
7 But ye shall die like men, and fall like one of the princes.
8 Arise, O God, judge the earth: for thou shalt inherit all nations.

Why do Zealots consider it Blasphemous to call yourself a God? Doesn't the Bible say that, didn't Jesus say that we are Gods? Why do people think calling themself a God, excludes them from having a creator? This verse says clearly that the children of the Most High are Gods!

Deuteronomy 8

All the commandments which I command thee this day shall ye observe to do, that ye may live, and multiply, and go in and possess the land which the Lord sware unto your fathers.

2 And thou shalt remember all the way which the Lord thy God led thee these forty years in the wilderness, to humble thee, and to prove thee, to know what was in thine heart, whether thou wouldest keep his commandments, or no.

3 And he humbled thee, and suffered thee to hunger, and fed thee with manna, which thou knewest not, neither did thy fathers know; that he might make thee know that man doth not live by bread only, but by every word that proceedeth out of the mouth of the Lord doth man live.

4 Thy raiment waxed not old upon thee, neither did thy foot swell, these forty years.

5 Thou shalt also consider in thine heart, that, as a man chasteneth his son, so the Lord thy God chasteneth thee.

6 Therefore thou shalt keep the commandments of the Lord thy God, to walk in his ways, and to fear him.

7 For the Lord thy God bringeth thee into a good land, a land of brooks of water, of fountains and depths that spring out of valleys and hills;

8 A land of wheat, and barley, and vines, and fig trees, and pomegranates; a land of oil olive, and

honey;

9 A land wherein thou shalt eat bread without scarceness, thou shalt not lack any thing in it; **a land whose stones are iron**, and out of whose hills thou mayest dig brass.

10 When thou hast eaten and art full, then thou shalt bless the Lord thy God for the good land which he hath given thee.

11 Beware that thou forget not the Lord thy God, in not keeping his commandments, and his judgments, and his statutes, which I command thee this day:

12 Lest when thou hast eaten and art full, and hast built goodly houses, and dwelt therein;

13 And when thy herds and thy flocks multiply, and thy silver and thy gold is multiplied, and all that thou hast is multiplied;

14 Then thine heart be lifted up, and thou forget the Lord thy God, which brought thee forth out of the land of Egypt, from the house of bondage;

15 Who led thee through that great and terrible wilderness, wherein were fiery serpents, and scorpions, and drought, where there was no water; who brought thee forth water out of the rock of flint;

16 Who fed thee in the wilderness with manna, which thy fathers knew not, that he might humble thee, and that he might prove thee, to do thee good at thy latter end;

17 And thou say in thine heart, My power and the might of mine hand hath gotten me this wealth.

18 But thou shalt remember the Lord thy God: for it is he that giveth thee power to get wealth, that he may establish his covenant which he sware unto thy fathers, as it is this day.

19 And it shall be, if thou do at all forget the Lord thy God, and walk after other gods, and serve them, and worship them, I testify against you this day that ye shall surely perish.

20 As the nations which the Lord destroyeth before your face, so shall ye perish; because ye would not be obedient unto the voice of the Lord your God.

I sincerely pray you are getting it, I hope you are buying these books by the dozen! The low-frequency bioelectrical signals may facilitate communication between different cells and tissues, impacting various physiological processes. For instance, cardiac cells communicate through electrical impulses, ensuring coordinated heartbeats. External electromagnetic fields, such as those from radiofrequency (RF) radiation, can interact with the body's natural low-frequency EMFs. This interaction might influence cellular functions, suggesting a more complex relationship between external electromagnetic exposure and biological responses.

Matthew 28

18 And Jesus came and spake unto them, saying, All power is given unto me in heaven and in earth.

19 Go ye therefore, and teach all nations, baptizing them in the name of the Father, and of the Son, and of the Holy Ghost:

20 Teaching them to observe all things whatsoever I have commanded you: and, lo, I am with you always, even unto the end of the world. Amen.

What have we been teaching? Plasma consists of moving electric fields, moving electric fields create magnetic fields, alternating electric and magnetic fields create light.

Electric

Magnetic

Light

Electric Fields generate Magnetic Fields

Magnetic Fields generate Electric Fields

The two generate Light, Light can make the two.

<u>NO CONTRADICTIONS WITHIN THIS TRINITY</u>.

Revelation 2

Unto the angel of the church of Ephesus write; These things saith he that holdeth the seven stars in his right hand, who walketh in the midst of the seven golden candlesticks;

2 I know thy works, and thy labour, and thy patience, and how thou canst not bear them which are evil: and thou hast tried them which say they are apostles, and are not, and hast found them liars:

3 And hast borne, and hast patience, and for my name's sake hast laboured, and hast not fainted.

4 Nevertheless I have somewhat against thee, because thou hast left thy first love.

5 Remember therefore from whence thou art fallen, and repent, and do the first works; or else I will come unto thee quickly, and will remove thy candlestick out of his place, except thou repent.

6 But this thou hast, that thou hatest the deeds of the Nicolaitanes, which I also hate.

7 He that hath an ear, let him hear what the Spirit saith unto the churches; To him that overcometh will I give to eat of the tree of life, which is in the midst of the paradise of God.

8 And unto the angel of the church in Smyrna write; These things saith the first and the last, which was dead, and is alive;

9 I know thy works, and tribulation, and poverty, (but thou art rich) and I know the blasphemy of them which say they are Jews, and are not, but are the synagogue of Satan.

10 Fear none of those things which thou shalt suffer: behold, the devil shall cast some of you into prison, that ye may be tried; and ye shall have tribulation ten days: be thou faithful unto death, and I will give thee a crown of life.

11 He that hath an ear, let him hear what the Spirit saith unto the churches; He that overcometh shall not be hurt of the second death.

12 And to the angel of the church in Pergamos write; These things saith he which hath the sharp sword with two edges;

13 I know thy works, and where thou dwellest, even where Satan's seat is: and thou holdest fast my name, and hast not denied my faith, even in those days wherein Antipas was my faithful martyr, who was slain among you, where Satan dwelleth.

14 But I have a few things against thee, because thou hast there them that hold the

doctrine of Balaam, who taught Balac to cast a stumblingblock before the children of Israel, to eat things sacrificed unto idols, and to commit fornication.

15 So hast thou also them that hold the doctrine of the Nicolaitanes, which thing I hate.

16 Repent; or else I will come unto thee quickly, and will fight against them with the sword of my mouth.

17 He that hath an ear, let him hear what the Spirit saith unto the churches; To him that overcometh will I give to eat of the hidden manna, and will give him a white stone, and in the stone a new name written, which no man knoweth saving he that receiveth it.

18 And unto the angel of the church in Thyatira write; These things saith the Son of God, who hath his eyes like unto a flame of fire, and his feet are like fine brass;

19 I know thy works, and charity, and service, and faith, and thy patience, and thy works; and the last to be more than the first.

20 Notwithstanding I have a few things against thee, because thou sufferest that woman Jezebel, which calleth herself a prophetess, to teach and to seduce my servants to commit fornication, and to eat things sacrificed unto idols.

21 And I gave her space to repent of her fornication; and she repented not.

22 Behold, I will cast her into a bed, and them that commit adultery with her into great tribulation, except they repent of their deeds.

23 And I will kill her children with death; and all the churches shall know that I am he which searcheth the reins and hearts: and I will give unto every one of you according to your works.

24 But unto you I say, and unto the rest in Thyatira, as many as have not this doctrine, and which have not known the depths of Satan, as they speak; I will put upon you none other burden.

25 But that which ye have already hold fast till I come.

26 And he that overcometh, and keepeth my works unto the end, to him will I give power over the nations:

27 And he shall rule them with a rod of iron; as the vessels of a potter shall they be broken to shivers: even as I received of my Father.

28 And I will give him the morning star.

29 He that hath an ear, let him hear what the Spirit saith unto the churches.

Revelation 3

And unto the angel of the church in Sardis write; These things saith he that hath the seven Spirits of God, and the seven stars; I know thy works, that thou hast a name that thou livest, and art dead.

2 Be watchful, and strengthen the things which remain, that are ready to die: for I have not found thy works perfect before God.

3 Remember therefore how thou hast received and heard, and hold fast, and repent. If therefore thou shalt not watch, I will come on thee as a thief, and thou shalt not know what hour I will come upon thee.

4 Thou hast a few names even in Sardis which have not defiled their garments; and they shall walk with me in white: for they are worthy.

5 He that overcometh, the same shall be clothed in white raiment; and I will not blot out his name out of the book of life, but I will confess his name before my Father, and before his angels.

6 He that hath an ear, let him hear what the Spirit saith unto the churches.

7 And to the angel of the church in Philadelphia write; These things saith he that is holy, he that is true, he that hath the key of David, he that

openeth, and no man shutteth; and shutteth, and no man openeth;

8 I know thy works: behold, I have set before thee an open door, and no man can shut it: for thou hast a little strength, and hast kept my word, and hast not denied my name.

9 **Behold, I will make them of the synagogue of Satan, which say they are Jews, and are not, but do lie; behold, I will make them to come and worship before thy feet, and to know that I have loved thee**.

10 Because thou hast kept the word of my patience, I also will keep thee from the hour of temptation, which shall come upon all the world, to try them that dwell upon the earth.

11 Behold, I come quickly: hold that fast which thou hast, that no man take thy crown.

12 Him that overcometh will I make a pillar in the temple of my God, and he shall go no more out: and I will write upon him the name of my God, and the name of the city of my God, which is new Jerusalem, which cometh down out of heaven from my God: and I will write upon him my new name.

13 He that hath an ear, let him hear what the Spirit saith unto the churches.

14 And unto the angel of the church of the Laodiceans write; These things saith the Amen, the faithful and true witness, the beginning of the creation of God;

15 I know thy works, that thou art neither cold nor hot: I would thou wert cold or hot.

16 So then because thou art lukewarm, and neither cold nor hot, I will spue thee out of my mouth.

17 Because thou sayest, I am rich, and increased with goods, and have need of nothing; and knowest not that thou art wretched, and miserable, and poor, and blind, and naked:

18 I counsel thee to buy of me gold tried in the fire, that thou mayest be rich; and white raiment, that thou mayest be clothed, and that the shame of thy nakedness do not appear; and anoint thine eyes with eyesalve, that thou mayest see.

19 As many as I love, I rebuke and chasten: be zealous therefore, and repent.

20 Behold, I stand at the door, and knock: if any man hear my voice, and open the door, I will come in to him, and will sup with him, and he with me.

21 To him that overcometh will I grant to sit with me in my throne, even as I also overcame,

and am set down with my Father in his throne.

22 He that hath an ear, let him hear what the Spirit saith unto the churches.

Beverly

Blavatsky

Hitler

Crowley

Parsons - Once they infiltrated Government and Science, they created a minstrel show to attract the Masses.

Anton LaVey & the Hollywood Weekend Satanists, just to popularize satanism with vampirism. It is a full court press, alternative news teaching Aliens aka demonology.

Demonology - the study of demons within religious belief and myth. Depending on context, it can refer to studies within theology, religious doctrine, or occultism. In many faiths, it concerns the study of a hierarchy of demons. Demons may be nonhuman separable souls, or discarnate spirits which have never inhabited a body. A sharp distinction is often drawn between these two classes, notably by the Melanesians, several African groups, and others. The Islamic jinn, for example, are not reducible to modified human souls. At the same time these classes are frequently conceived as producing identical results, e.g. diseases.

Angels & Demons - Dan Brown, remixed from the other side of the Bible Daemonologie. King James not only wrote the Bible he wrote a book on Demons.

The twisting of truths is the Devil's work. I mean search the internet, I can't get over the amount of Bible Scholars flaunting their degrees as if...

The University of Bologna in Italy, regarded as the oldest university in Europe, was the first institution to confer the degree of Doctor in Civil Law in the late 12th century; it also conferred similar degrees in other subjects, including medicine. The history of higher education in the United States begins in 1636 and continues to the present time. American higher education is known throughout the world for its dramatic expansion. It was also heavily influenced by British models in the colonial era, and German models in the 19th century. The American model includes private schools, mostly founded by religious denominations, as well as universities run by state governments, and a few military academies that are run by the national government. The Bible predates Degrees, the Bible is for every man, woman and child. In the United States, departments of Religious Studies began to emerge in public universities beginning in the late 1950s and 1960s. Originally you studied alone and with family, then after showing a inclination, you were sent to study with people in the movement. Very similar to a trade, a master and an apprentice. Start slapping these charlatans up, when they introduce themselves with their Degrees. These degree guys lead to this mess...

Do you know what burned brass looks like? Bwahahahahaha...

Revelation 1

13 And in the midst of the seven candlesticks one like unto the Son of man, clothed with a garment down to the foot, and girt about the paps with a golden girdle.
14 His head and his hairs were white like wool, as white as snow; and his eyes were as a flame of fire;
15 And his feet like unto fine brass, as if they burned in a furnace; and his voice as the sound of many waters.

Burned Brass is black black, Eumelanin. This face here is a white man, that is not even fine brass colored, I am fine brass colored! The picture of Jesus never matters until you make one Black...
I mean communion is eating the flesh of Christ and drinking the blood of Christ, right? How doesn't it matter? If you ask me, I will say that as a man he is EuMelanin Dominant period, as Light he is White. I just think to over look definitive passages in the Bible or "Fake Degree" them up... is crazy!
The Word is Flesh, that can only be EuMelanin.
Being 'Black' is not a curse, it is a Blessing. Being ignorant is a curse, not a blessing.

Danniel 7

In the first year of Belshazzar king of Babylon

Daniel had a dream and visions of his head upon his bed: then he wrote the dream, and told the sum of the matters.

2 Daniel spake and said, I saw in my vision by night, and, behold, **the four winds of the heaven** strove upon the great sea.

3 And **four great beasts came up from the sea**, diverse one from another.

4 The first was like a lion, and had eagle's wings: I beheld till the wings thereof were plucked, and it was lifted up from the earth, and made stand upon the feet as a man, and a man's heart was given to it.

5 And behold another beast, a second, like to a bear, and it raised up itself on one side, and it had three ribs in the mouth of it between the teeth of it: and they said thus unto it, Arise, devour much flesh.

6 After this I beheld, and lo another, like a leopard, which had upon the back of it four wings of a fowl; the beast had also four heads; and dominion was given to it.

7 After this I saw in the night visions, and behold a fourth beast, dreadful and terrible, and strong exceedingly; and it had great iron teeth: it devoured and brake in pieces, and stamped the residue with the feet of it: and it was diverse from all the beasts that were before it; **and it had ten horns**.

8 **I considered the horns, and, behold, there came up among them** <u>another little horn</u>,

before whom there were three of the first horns plucked up by the roots: and, behold, in this horn were eyes like the eyes of man, and a mouth speaking great things.

9 I beheld till the thrones were cast down, and the Ancient of days did sit, whose garment was white as snow, and the hair of his head like the pure wool: his throne was like the fiery flame, and his wheels as burning fire.

10 A fiery stream issued and came forth from before him: thousand thousands ministered unto him, and ten thousand times ten thousand stood before him: the judgment was set, and the books were opened.

11 <u>**I beheld then because of the voice of the great words which the horn spake**</u>: I beheld even till the beast was slain, and his body destroyed, and given to the burning flame.

12 As concerning the rest of the beasts, they had their dominion taken away: yet their lives were prolonged for a season and time.

13 I saw in the night visions, and, behold, one like the Son of man came with the clouds of heaven, and came to the Ancient of days, and they brought him near before him.

14 And there was given him dominion, and glory, and a kingdom, that all people, nations, and languages, should serve him: his dominion is an everlasting dominion, which shall not pass away, and his kingdom that which shall not be destroyed.

15 I Daniel was grieved in my spirit in the midst of my body, and the visions of my head troubled me.

16 I came near unto one of them that stood by, and asked him the truth of all this. So he told me, and made me know the interpretation of the things.

17 **<u>These great beasts, which are four, are four kings, which shall arise out of the earth</u>**.

18 <u>But the saints of the most High shall take the kingdom, and possess the kingdom for ever, even for ever and ever</u>.

19 Then I would know the truth of the fourth beast, which was diverse from all the others, exceeding dreadful, whose teeth were of iron, and his nails of brass; which devoured, brake in pieces, and stamped the residue with his feet;

20 **<u>And of the ten horns that were in his head, and of the other which came up, and before whom three fell; even of that horn that had eyes, and a mouth that spake very great things, whose look was more stout than his fellows</u>**.

21 **<u>I beheld, and the same horn made war with the saints, and prevailed against them</u>**;

22 Until the Ancient of days came, and judgment was given to the saints of the most High; and the time came that the saints possessed the kingdom.

23 Thus he said, The fourth beast shall be the fourth kingdom upon earth, which shall be diverse from all kingdoms, and shall devour the

whole earth, and shall tread it down, and break it in pieces.

24 <u>And the ten horns out of this kingdom are ten kings that shall arise: and another shall rise after them; and he shall be diverse from the first, and he shall subdue three kings</u>.

25 And he shall speak great words against the most High, and shall wear out the saints of the most High, and think to change times and laws: and they shall be given into his hand until a time and times and the dividing of time.

26 But the judgment shall sit, and they shall take away his dominion, to consume and to destroy it unto the end.

27 And the kingdom and dominion, and the greatness of the kingdom under the whole heaven, shall be given to the people of the saints of the most High, whose kingdom is an everlasting kingdom, and all dominions shall serve and obey him.

28 Hitherto is the end of the matter. As for me Daniel, my cogitations much troubled me, and my countenance changed in me: but I kept the matter in my heart.

Zechariah 1

In the eighth month, in the second year of Darius, came the word of the Lord unto Zechariah, the son of Berechiah, the son of Iddo the prophet, saying,

2 The Lord hath been sore displeased with **your fathers**.

3 Therefore say thou unto them, Thus saith the Lord of hosts; Turn ye unto me, saith the Lord of hosts, and I will turn unto you, saith the Lord of hosts.

4 Be ye not as your **fathers**, **unto whom the former prophets have cried**, saying, Thus saith the Lord of hosts; Turn ye now from your evil ways, and from your evil doings: but they did not hear, nor hearken unto me, saith the Lord.

5 Your fathers, where are they? and the prophets, do they live for ever?

6 <u>But my words and my statutes, which I commanded my servants the prophets, did they not take hold of your fathers</u>? and they returned and said, Like as the Lord of hosts thought to do unto us, according to our ways, and according to our doings, so hath he dealt with us.

7 Upon the four and twentieth day of the eleventh month, which is the month Sebat, in the second year of Darius, came the word of the Lord unto Zechariah, the son of Berechiah, the son of Iddo the prophet, saying,

8 I saw by night, and behold a man riding upon a red horse, and he stood among the myrtle trees that were in the bottom; and behind him were there red horses, speckled, and white.

9 Then said I, O my lord, what are these? And the angel that talked with me said unto me, I will shew thee what these be.

10 And the man that stood among the myrtle trees answered and said, These are they whom the Lord hath sent to walk to and fro through the earth.

11 And they answered the angel of the Lord that stood among the myrtle trees, and said, We have walked to and fro through the earth, and, behold, all the earth sitteth still, and is at rest.

12 Then the angel of the Lord answered and said, O Lord of hosts, how long wilt thou not have mercy on Jerusalem and on the cities of Judah, against which thou hast had indignation these threescore and ten years?

13 And the Lord answered the angel that talked with me with good words and comfortable words.

14 So the angel that communed with me said unto me, Cry thou, saying, Thus saith the Lord of hosts; I am jealous for Jerusalem and for Zion with a great jealousy.

15 <u>And I am very sore displeased with the heathen that are at ease</u>: for I was but a little displeased, and they helped forward the affliction.

16 Therefore thus saith the Lord; I am returned to Jerusalem with mercies: my house shall be built in it, saith the Lord of hosts, and a line shall be stretched forth upon Jerusalem.

17 Cry yet, saying, Thus saith the Lord of hosts; My cities through prosperity shall yet be spread abroad; and the Lord shall yet comfort Zion, and

shall yet choose Jerusalem.

18 **Then lifted I up mine eyes, and saw, and behold four horns**.

19 And I said unto the angel that talked with me, What be these? And he answered me, **These are the horns which have scattered Judah, Israel, and Jerusalem**.

20 And **the Lord shewed me four carpenters**.

21 Then said I, What come these to do? And he spake, saying, **These are the horns which have scattered Judah, so that no man did lift up his head: but these are come to fray them, to cast out the horns of the Gentiles, which lifted up their horn over the land of Judah to scatter it**.

The Four Naggars are going to... WHAT!!!

Quran 9:30 (At-Tawbah)

The Jews say, "Uzayr is the son of Allah," while the Christians say, "The Messiah is the son of Allah." Such are their baseless assertions, only parroting the words of earlier disbelievers. May Allah condemn them! How can they be deluded ⌐from the truth¬?

Ezekiel 29: 6

And all the inhabitants of Egypt shall know that I am the Lord, because they have been a staff of reed to the house of Israel.

The Elephantine Paryri is Aramaic, and may agree with the Quran, that Wusir was worshipped by the Jews. Wusir, Ausar, Osiris is still the God of Biology you know, he is just secretly referred to via EuMelanin. The Word that became Flesh, is still under a close Microscope.

The Carpentras Stele is a stele found at Carpentras in southern France in 1704 that contains the first published inscription written in the Phoenician alphabet, and the first ever identified (a century later) as Aramaic. It remains in Carpentras, at the Bibliothèque Inguimbertine, in a "dark corner" on the first floor. Older Aramaic texts were found since the 9th century BC, but this one is the first Aramaic text to be published in Europe. It is known as KAI 269, CIS II 141 and TAD C20.5.

It is a funerary dedication to an unknown lady called Taba; the first line of the image depicts her standing before the god of the underworld with her arms raised and the second, her lying down, dead, being prepared for burial. The textual inscription is typical of Egyptian funerary tablets in that she is described as having done nothing bad in her life, and wishes her well in the presence of Osiris. Some of the words on the stele corresponded to the Aramaic in the <u>Book of Daniel</u>, and in the **Book of Ruth**.

Wait the Carpenter Stele? Naggar!

Ptolemy Philadelphus in the 3rd Century says the Bible is the new history of Egypt, allegedly. Egypt's true history rewritten, to mask those that were conquered. By these standards, you could say that all 3 modern religions were launched at Mouseion. Yes, extrapolate all you want, because it is **spooky** how close Mount Zion sounds to Mouseion. We must consider the Ptolemaic Ahmehstrahans and The Priests of Moses. The Ptolemaic Ahmehstrahans and The Priests of Moses were here in the Library of Alexandria. This is where Masonry began, stolen and counterfeited by Euclid. Mouseion was a huge "Campus", this is the true Origin of Religious Degrees. The original Religious Degrees are what you people call Masonic Degrees, only it wasn't religion like today's religion, it was akin to today's science. In fact the emergence of the 72 scholars in debate circles is creating some remarkably thought provoking insights. I have already told you my stance, I am comfortable with the Perfect Black Flesh, radiating pure White Light, as long as the science is intact, I have no beef. You see the 72 scholars is still a reference to Light. Modern Bible Scholars tore into the Letter of Aristeas to Philocrates. They are mad basically because the letter makes the Jews saviors to the Greeks/Romans aka Gentiles. The problem with this is, the Bible echoes these sentiments. That Israel should teach, will teach, will be a light to, and light for, the Gentiles. Under oppression is when people use code, and the 72 scholars, writing the text in 72 days...

The Letter of Aristeas to Philocrates is a **Hellenistic** work of the 3rd or early 2nd century BC, considered by some Biblical scholars to be **pseudepigraphical**. The letter is the earliest text to mention the **Library of Alexandria**.

Josephus, who paraphrases about two-fifths of the letter, ascribes it to **Aristeas of Marmora** and to have been written to a certain Philocrates. The letter describes the Greek translation of the **Hebrew Bible** by seventy-two interpreters sent into **Egypt** from **Jerusalem** at the request of the librarian of **Alexandria**, resulting in the **Septuagint** translation. Some scholars have since argued that it is fictitious.

I said I wasn't but I must... I need you to understand the essence of Magickal Ritual, and the reason all culture have rituals. It is hidden in the word Naggar.

Naggar:

Kmet & Kmer were the ruling elite of these Naga. Why do people that love the Legacy of Ancient Egypt or Ancient Kmt, hate being called Naggar! How Swaye? That is mental illness, cognitive dissonance on steroids!

There are Black Freemasons and Eastern Stars that hate being called Naggar! How Swaye? That is mental illness, cognitive dissonance on steroids! The Naga people are believed by some to be an ancient tribe who once inhabited Sri Lanka and various parts of Southern India. There are references to Nagas in several ancient texts such as Mahavamsa, Manimekalai, Mahabharata and also in
other Sanskrit and Pali literature. They were generally represented as a class of super-humans taking the form of serpents who inhabit a subterranean world. - wiki

There are Black Christians that love Jesus and hate being called Naggar! How Swaye? That is mental illness, cognitive dissonance on steroids!

The bible calls Jesus a Naggar.

However, the Greek term tektōn does not carry

this (bible) meaning, and the nearest equivalent in the New Testament is Paul's comparison to Timothy of a "workman" (ἐργάτης, ergatēs) **rightly "dividing" the word of truth**. This has been taken as carpentry by some Christian commentators. The suggested term naggar ("craftsman") is not found in biblical Aramaic or Hebrew, or in Aramaic documents of the New Testament period, but is found in later Talmudic texts where the term "craftsman" is used as a metaphor for **a skilled handler of the word of God**. - wiki

For more on this read &/or. reread the L'Goat book...

The interesting thing is it is similar to Ecclesiastes, meaning teacher or public speaker. Time, Space, Light & Sound are created in your mind. A ritual is designed to create a future or to re-member a past event. This is why a great many Altars are built in the build to Co-Memmorate a special event. Every time you re-count the event, you breath life into it, for current standing.

This is the reason the Math in the Kings Chamber is important, it is the same Math surrounding Jesus's 6th day, 3rd day (72 hour), in the 9th hour event. The King's Chamber is a empty tomb in a mountain, I don't understand all the confusion except for Racism and Power. Which I can understand from the Gentile perspective, but amongst the children? Christ is the Light of the World, it doesn't matter if you say that in English, Greek, Hebrew, Aramaic or Ancient Kemetic tongues... We are getting lost in debates about God and not communing with God. We are getting lost in debates about God and not following God. God does need us to strengthen him though, this is a hard pill to swallow for many. Faith is required not just for us, but for him. The field, the waveguide is being polluted, with sex, money and murder. Yes, it is to our benefit and the benefit of the planet, the creator won't die but his children... We are entangled, making us suffer is the only way the Devil has to make him suffer. This is why at the end of the day it truly just comes down to taking good men and making them better. This produces better PTathers, betters sons and daughters, conserving the ideal of Morality, Uprightness, Righteousness. Liberalism built from warped Meliorism & Individualism (Humanism), is intellectual Satanism. I don't think John Locke was what we term a Liberalist today even though he never was into women... I will leave that alone (I am probably out of my league in political history waters).

Back to Work!

Religious scholars are saying that the enslavement of the Herbews by the Egyptians, is actually the enslavement of the Egyptians by the Greeks. Scholars are saying this was done carefully to preserve their history, while treading very lightly in such close proximity to their oppressors. This make sense when you get to the nitty gritty of chronology, and ask for the original Hebrew version of the Bible. You will be redirected to the Septuagint. The Septuagint was written in Greek, by Hebrew first speaking Jews. The story is true though, the Jews were enslaved by the Egyptians. The thing is time and space, the time and space was changed. The Jews were those original Egyptians, the Egyptians were the old Greeks. The Old Greeks became the Egyptians in real life when they enslaved the Egyptians. They then rewrote all their books, scrolls and papyri, not all but a good bit! Allegedly.

The word made flesh is the secret to the Bible, to the Biblical History, Jesus and Salvation. The Word made Flesh is EuMelanin (it's distant cousin Chlorophyll too), we can understand this easily. Especially if you have been diligently studying all these books... You did buy this book from Amazon right? Did you see the name of the series this book belongs to?

Discussing the Neteru happens everyday, in every class and conversation about science. The Periodic table is different types of Atums, Atum is the word that became flesh. This is well understood by the authors of the Bible. Atum, Atom and Adam must be and stay intact, Adam is about the first creation of Radio Waves, Infrared Light. The Neters are energies, not chakras or shocks of ra either... Bwahahahahaha... Adam was in the Garden, that is historically known as the Nile Valley. Leaving the lush equatorial environment where food grows by itself created the hunters, to compete with the people that were native to cold climates. The hunter/gatherer never existed, only hunter and gatherers.

Gatherers - Equatorial Haplotype People

Hunters - Polar Haplotype People

Think! You can't gather food, where food doesn't grow. There is no need to hunt where food grows abundantly on it's own (God grown). **<u>The Yod, which was my favorite letter in Ivrit, scholars have now identified as the Shut or feather of Shu.</u>** That makes sense too, I will show you why later! This is a big secret, the Greeks had to totally violate Horus, they turned him into a boy with his finger over his lips. He guards the secret, ssshhhhhhh... That is the secret though, hidden in plain sight. Shu, the waveguide, the waveguide is the Living Record. The Living Record is Spinning and Carbon is the secret!

Shellac - Amber type Resin secreted by female lac bugs.

Vinyl - Poisonous PVC, yep them old records

are toxic! Do any DJs know this? PVC has toxic Phthalates & Lead in them, Bwahahahahaha... I told yall rap was genocide carefully crafted by the Germans. This is high level Biological Warfare...

Think about your favorite old skool DJ now, licking his fingers cutting the records!!! He had no clue, I had no clue, I was licking Lead and Phthalates!!!

Again here is the definition of Rap and the key to understanding all the shooting, charges and indictments.

Rap - mid-14c., rappen, "to strike, smite, knock," from rap (n.). Related: Rapped; rapping. To rap (someone's) knuckles "give sharp punishment" is from 1749 (to rap (someone's) fingers in the same sense is by 1670s.). Related: Rapped; rapping.
early 14c., rappe, "a quick, light blow; a resounding stroke," also "a fart" (late 15c.), native or borrowed from a Scandinavian source (compare Danish rap, Swedish rapp "light blow"); either way probably of imitative origin (compare slap, clap).
Slang meaning "**a rebuke**, the **blame**, responsibility" is from 1777; specific meaning "**<u>criminal indictment</u>**" (as in **rap sheet**, 1960) is from 1903; to beat the rap is from 1927. Meaning "music with improvised words" was in New York

City slang by 1979 (see rap (v.2)).

Back to History, why do you think the Jews hid amongst and married into the Egyptians? They were one and the same people genetically! Another reason they ate the mummies, they were hiding evidence. Its almost like the Homosexual Kiss of Denzel covering up the Racial Dynamic in Rome/Greece. The Gladiator 2 movie creates some interesting questions. What were the Hellenistic Jews studying? Wait do you see the need to encode the Bible yet? The authors were under oppression! Allegedly.

Acts 7

Then said the high priest, Are these things so?

2 And he said, Men, brethren, and fathers, hearken; The God of glory appeared unto our father Abraham, when he was in Mesopotamia, before he dwelt in Charran,

3 And said unto him, Get thee out of thy country, and from thy kindred, and come into the land which I shall shew thee.

4 Then came he out of the land of the Chaldaeans, and dwelt in Charran: and from thence, when his father was dead, he removed him into this land, wherein ye now dwell.

5 And he gave him none inheritance in it, no, not

so much as to set his foot on: yet he promised that he would give it to him for a possession, and to his seed after him, when as yet he had no child.

6 And God spake on this wise, That his seed should sojourn in a strange land; and that they should bring them into bondage, and entreat them evil four hundred years.

7 And the nation to whom they shall be in bondage will I judge, said God: and after that shall they come forth, and serve me in this place.

8 And he gave him the covenant of circumcision: and so Abraham begat Isaac, and circumcised him the eighth day; and Isaac begat Jacob; and Jacob begat the twelve patriarchs.

9 And the patriarchs, moved with envy, sold Joseph into Egypt: but God was with him,

10 And delivered him out of all his afflictions, and gave him favour and wisdom in the sight of Pharaoh king of Egypt; and he made him governor over Egypt and all his house.

11 Now there came a dearth over all the land of Egypt and Chanaan, and great affliction: and our fathers found no sustenance.

12 But when Jacob heard that there was corn in Egypt, he sent out our fathers first.

13 And at the second time Joseph was made known to his brethren; and Joseph's kindred was made known unto Pharaoh.

14 Then sent Joseph, and called his father Jacob to him, and all his kindred, threescore and fifteen souls.

15 So Jacob went down into Egypt, and died, he, and our fathers,

16 And were carried over into Sychem, and laid in the sepulchre that Abraham bought for a sum of money of the sons of Emmor the father of Sychem.

17 But when the time of the promise drew nigh, which God had sworn to Abraham, the people grew and multiplied in Egypt,

18 Till another king arose, which knew not Joseph.

19 The same dealt subtilly with our kindred, and evil entreated our fathers, so that they cast out their young children, to the end they might not live.

20 In which time Moses was born, and was exceeding fair, and nourished up in his father's house three months:

21 And when he was cast out, Pharaoh's

daughter took him up, and nourished him for her own son.

22 **And Moses was learned in all the wisdom of the Egyptians, and was mighty in words and in deeds.**

Exodus 2

11 And it came to pass in those days, when Moses was grown, that he went out unto his brethren, and looked on their burdens: and he spied an Egyptian smiting an Hebrew, one of his brethren.

12 And he looked this way and that way, and when he saw that there was no man, he slew the Egyptian, and hid him in the sand.

13 And when he went out the second day, behold, two men of the Hebrews strove together: and he said to him that did the wrong, Wherefore smitest thou thy fellow?

14 And he said, Who made thee a prince and a judge over us? intendest thou to kill me, as thou killedst the Egyptian? And Moses feared, and said, Surely this thing is known.

15 Now when Pharaoh heard this thing, he sought to slay Moses. But Moses fled from the face of Pharaoh, and dwelt in the land of Midian: and he sat down by a well.

16 Now the priest of Midian had seven daughters: and they came and drew water, and filled the troughs to water their father's flock.

17 And the shepherds came and drove them away: but Moses stood up and helped them, and watered their flock.

18 And when they came to Reuel their father, he said, How is it that ye are come so soon to day?

19 And they said, An Egyptian delivered us out of the hand of the shepherds, and also drew water enough for us, and watered the flock.

20 And he said unto his daughters, And where is he? why is it that ye have left the man? call him, that he may eat bread.

21 And Moses was content to dwell with the man: and he gave Moses Zipporah his daughter.

22 And she bare him a son, and he called his name Gershom: for he said, I have been a stranger in a strange land.

23 And it came to pass in process of time, that the king of Egypt died: and the children of Israel sighed by reason of the bondage, and they cried, and their cry came up unto God by reason of the bondage.

24 And God heard their groaning, and God

remembered his covenant with Abraham, with Isaac, and with Jacob.

25 And God looked upon the children of Israel, and God had respect unto them.

OK, key points for this era, not only was Moses the Law Giver educated in the Mysteries of Osiris & Atum, it follows that he was mighty in words and deeds. This implies he was mighty in words and deeds because of his education. Then we can't overlook the fact that the Israelites were worshipping Apis (son of Hathor)! I don't see why everyone isn't a Christian! Did we skip the moment David crowned himself, keep in mind there is only 1 crowning event in the Bible!

2 Samuel 12

And the Lord sent Nathan unto David. And he came unto him, and said unto him, There were two men in one city; the one rich, and the other poor.

2 The rich man had exceeding many flocks and herds:

3 But the poor man had nothing, save one little ewe lamb, which he had bought and nourished up: and it grew up together with him, and with his children; it did eat of his own meat, and drank of his own cup, and lay in his bosom, and was unto him as a daughter.

4 And there came a traveller unto the rich man, and he spared to take of his own flock and of his own herd, to dress for the wayfaring man that was come unto him; but took the poor man's lamb, and dressed it for the man that was come to him.

5 And David's anger was greatly kindled against the man; and he said to Nathan, As the Lord liveth, the man that hath done this thing shall surely die:

6 And he shall restore the lamb fourfold, because he did this thing, and because he had no pity.

7 And Nathan said to David, Thou art the man. Thus saith the Lord God of Israel, I anointed thee king over Israel, and I delivered thee out of the hand of Saul;

8 And I gave thee thy master's house, and thy master's wives into thy bosom, and gave thee the house of Israel and of Judah; and if that had been too little, I would moreover have given unto thee such and such things.

9 Wherefore hast thou despised the commandment of the Lord, to do evil in his sight? thou hast killed Uriah the Hittite with the sword, and hast taken his wife to be thy wife, and hast slain him with the sword of the children of Ammon.

10 Now therefore the sword shall never depart from thine house; because thou hast despised me, and hast taken the wife of Uriah the Hittite to be thy wife.

11 Thus saith the Lord, Behold, I will raise up evil against thee out of thine own house, and I will take thy wives before thine eyes, and give them unto thy neighbour, and he shall lie with thy wives in the sight of this sun.

12 For thou didst it secretly: but I will do this thing before all Israel, and before the sun.

13 And David said unto Nathan, I have sinned against the Lord. And Nathan said unto David, The Lord also hath put away thy sin; thou shalt not die.

14 Howbeit, because by this deed thou hast given great occasion to the enemies of the Lord to blaspheme, the child also that is born unto thee shall surely die.

15 And Nathan departed unto his house. And the Lord struck the child that Uriah's wife bare unto David, and it was very sick.

16 David therefore besought God for the child; and David fasted, and went in, and lay all night upon the earth.

17 And the elders of his house arose, and went to

him, to raise him up from the earth: but he would not, neither did he eat bread with them.

18 And it came to pass on the seventh day, that the child died. And the servants of David feared to tell him that the child was dead: for they said, Behold, while the child was yet alive, we spake unto him, and he would not hearken unto our voice: how will he then vex himself, if we tell him that the child is dead?

19 But when David saw that his servants whispered, David perceived that the child was dead: therefore David said unto his servants, Is the child dead? And they said, He is dead.

20 Then David arose from the earth, and washed, and anointed himself, and changed his apparel, and came into the house of the Lord, and worshipped: then he came to his own house; and when he required, they set bread before him, and he did eat.

21 Then said his servants unto him, What thing is this that thou hast done? thou didst fast and weep for the child, while it was alive; but when the child was dead, thou didst rise and eat bread.

22 And he said, While the child was yet alive, I fasted and wept: for I said, Who can tell whether God will be gracious to me, that the child may live?

23 But now he is dead, wherefore should I fast? can I bring him back again? I shall go to him, but he shall not return to me.

24 And David comforted Bathsheba his wife, and went in unto her, and lay with her: and she bare a son, and he called his name Solomon: and the Lord loved him.

25 And he sent by the hand of Nathan the prophet; and he called his name Jedidiah, because of the Lord.

26 **And Joab fought against Rabbah of the children of Ammon, and took the royal city**.

27 And Joab sent messengers to David, and said, I have fought against Rabbah, and have taken the city of waters.

28 Now therefore gather the rest of the people together, and encamp against the city, and take it: lest I take the city, and it be called after my name.

29 And David gathered all the people together, and went to Rabbah, and fought against it, and took it.

30 **And he took their king's crown from off his head, the weight whereof was a talent of gold with the precious stones: and it was set on David's head. And he brought forth the spoil of**

the city in great abundance.

31 And he brought forth the people that were therein, and put them under saws, and under harrows of iron, and under axes of iron, and made them pass through the brick-kiln: and thus did he unto all the cities of the children of Ammon. So David and all the people returned unto Jerusalem.

What was this crown? I already told you the TopHat preserved in Masonry is Osiris's crown, what is Ammon known for? **The horns of Ammon** were curling ram horns, used as a symbol of the Egyptian deity Ammon (also spelled Amun or Amon). Because of the visual similarity, they were also associated with the fossils shells of ancient snails and cephalopods, the latter now known as ammonite because of that historical connection. **The horns of Ammon** is what they crowned David with, interesting huh? I mean it's that or the Serpent lmao... The serpent we know represents Light and the Spine/Brain. This is all biochemistry, the science of Life!

Shekhinah (Hebrew: שְׁכִינָה, Modern: Šəḵīna, Tiberian: Šeḵīnā) is the English transliteration of a Hebrew word meaning "dwelling" or "settling" and denotes the presence of God in a place. This concept is found in Judaism and the Torah, as mentioned in Exodus 25:8.

The word "Shekhinah" is not found in the Bible. It appears in the Mishnah, the Talmud, and Midrash. The baby Horus is Light, in a Cell. Conception, the dwelling place of God is the Human Cell, science also says that electromagnetism begins in Human Cells, 0 hertz to ____.

Circumcision is key to being a true Jew correct? I already discussed the divinity in the 8th day etc.... Please read our previous books, but a picture is worth a thousand words right?

Keep in mind what makes this book and me different, I am saying we must stop the Racism. There was never a Israel vs Egypt, just a Egypt/Israel vs Greece/Rome.

It was actually the last King of Israel that is credited with finding the Law of Moses, this means that the children of Israel like the Quran says, was following the Mysteries of Wuzir. Josiah said none of the Kings of Israel followed Moses Law. The Quran goes even further to say that Allah returned Egypt to the Jews, it's hotly debated now!

OCCUPATION

Occupation - early 14c., "**fact of holding or possessing**;" mid-14c., "a being employed in something," also "a particular action," from Old French occupacion "pursuit, work, employment; occupancy, occupation" (12c.), from Latin occupationem (nominative occupatio) "a taking possession; business, employment," noun of action from past-participle stem of occupare (see occupy). Meaning "employment, business in which one engages" is late 14c. That of "condition of being held and ruled by troops of another country" is from 1940.

Athlete - early 15c. (Chauliac), "competitor in athletic games and contests," from Latin athleta "a wrestler, athlete, combatant in public games," from Greek athlētēs "prizefighter, contestant in the games," agent noun from athlein "to contest for a prize," which from athlos "a contest," especially for a prize (its neuter form, athlon, meant "the prize of a contest"), a word of unknown origin.
Until mid-18c. it was usually in English in

Latin form. Old English had plegmann "**play-man**." The meaning "one trained in exercises of agility and strength" is by 1827. Athlete's foot is recorded by 1928, for an ailment that has been around much longer.

Why are athletes getting paid more and more? Better question, what does an athlete do that increases in value? Wait... A basketball player, scores, stops scores, why does their pay increase? Their pay increases because the money the leagues get increases, so they are still under paid. Yeah... If you are thinking that was somewhere in the right ball park, your wrong. I am going to say that answer is correct though, and I probably asked a bad question. Why are the leagues of major sports making more money? Commercial companies are paying much higher prices to advertise. Yeah... If you are thinking that was somewhere in the right ball park, your wrong. I am going to say that answer is correct though, and I probably asked a bad question. What does the sports leagues sell to the Commercial Companies? AirTime! Booom! We done. Close this book!

You don't get it? First of all air is free, but nah... It's your attention, they harvest, package and sale your attention! Your attention as the children of Israel is valuable too... You are the **plug**. You are **paying** attention... You are **spending** time, **paying** attention. Sports leagues

leverage your **interest**!!! They know the truth is coming, the closer we get to the DEADLINE, the more important your attention becomes, the more valuable distractions become.

Athletes occupation like all entertainers **Duty**, **Job** is to Occupy your Time. Time equals Money, more time you **spend**, **paying** attention to them, the less of a threat you are. Satan is winning.

Real Estate is a good business because the people keep reproducing, as the population grows there is less space. The space decreases causes the value of the space to increase. Every year though, there are more and more young athletes, making the uniqueness decrease. This why it's so hard for athletes to sound intelligent when they speak, they are relatively dumb. Here is a sure fire sign a athlete is dumb:

Interviewer: Anything you wanna say or teach the next generation about?

Seasoned Athlete: Generational Wealth.

Interviewer: How did you learn about Generational Wealth?

Seasoned Athlete: One of my **owners** really enlightened me. He has really **mastered** the art of making money. He is a Genius!

Dummies! The owners don't make money, they just have relationships. They have agreed to your enslavement. You will be owned by them for X-Time and during that time, you will pick Almonds. You will pick Almonds for all the other Masters they have relationships with. Then they will instruct you on which of their relationships you should later invest in, maybe even partner with them... just like Kanye & Puffy! You should be teaching your children about Generational Health! That's how Tom Brad and Lebron James mastered their sports, being the healthiest! The longer they are 'live' the more views they get... Let me get back to the Line of Shudah, I mean the loins of Shu, this darn keyboard, I mean the Lion of Judah.

I think Exodus is where we should start, Exodus is a fake name, better still a code name. The code is we need to escape occupation, they have our minds hostage and our Intellectual Property. They are making us rewrite history so we can worship them, as us. Get it? The Bible is not plagiarized, it is written by the True Egyptians. They left bread crumbs:

Impuwer Papyrus Blood is Everywhere

Exodus 7:21 there was blood throughout the land of Egypt

Impuwer Papyrus Men shrink from human beings and thirst after water

Exodus 7:24 and all the Egyptians rigged round

about the river for water to drink

Impuwer Papyrus he who places his brother in the ground is everywhere

Exodus 12:30 there was not a house where there was not one dead

You were lead to believe you needed to be at war, Israelite vs Egyptian. I never subscribed to that, I have always maintained this line of thought, the Egyptians are the Israelites are the Tribe of Judah is the Tribe of Light. I mean we can just all go stand in the sun and test the theory the easy way...

EuMelanin = The Perfect Black, the original Green Screen!

Exodus right? Yall know I get distracted so easy. Exodus is not in the Torah, Exodus is only in English translations.

SH'MOT

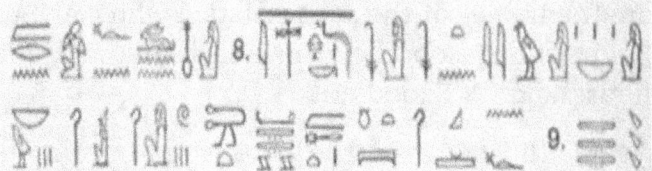

8. Homage to thee, King of Kings, and Lord of Lords, and Prince of Princes. Thou hast ruled the Two Lands from the womb of the goddess Nut.⁶ Thou hast governed the 9. Lands of Akert.⁷ Thy members are of silver-

In case this doesn't print well line 8 says Homage to thee, King of Kings and Lord of Lords....

The real name of Exodus is Sh'Mot, yep. It's Sh'Mot identifying as Exodus!

We simmering the food relax...

Revelation 19:16

And he hath on his vesture and on his thigh a name written, King Of Kings, And Lord Of Lords.

Sh'Mot - I did a quick google and found this from. FirmIsrael.com: The Meaning of "Exodus"; The name of the second book, Exodus, gets a little more complicated. It is a Greek word that means "going out" and points to the content of

Israel leaving Egypt. However, as mentioned, this book was not originally written in Greek. So Exodus is not its original name! To Hebrew readers, this is the book of Sh'mot, which means "names", as found in the first verse: "Now these are the names of the sons of Israel who came to Egypt with Jacob; they came each one with his household..." -Exodus 1:1

Again, the name of the book comes from its very first verse. To the ancient Hebrew mind, a name was not just a unique or identifying title of a person, but spoke of someone's character or destiny. Moses means drawn out — His life was used to draw Israel out of Egypt. Elijah means The Lord is my God, and His life was marked by bold and uncompromising stands for the Lord. And likely the most important of them all – Jesus means salvation. Names have valuable meaning. Jesus is not there either though... You see what I am saying, codes inside codes. The name would've been something like Shu...

The children of Egypt, were known as the children of Shu. The Sphinx is about Shu the Lion and Anubis the Rottweiler.

Fun Fact: You were also lied to about Abraham's origins, Ur is not for Chaldea, Ur is for Egypt! Ur is the one word that means Lion and Dog in Sumerian. When they wrote Abraham was from Ur it was code, meaning a BlackMan (Ham) from the Land of the Great Lion &/or Great Dog.

Back to Jesus, Yashua or Yehoshua would've been the actual name, this translates to Joshua. Do you see how Shu is in there? They even spun the gentiles with other Joshuas, they didn't want to get caught and put to death. This is how you get Jesus Christ as Yeshua HaMashiach the Anointed Salvation bringer.

In Kemetic Science who is the mother of Shu? The Holy Ghost, ok I made that part up... Bwahahahaha... We was cooking though lmao... Atum or the Hidden one was in the Plasma Soup aka Nun swimming.

Stop, that's a easy lay up, sperm swimming in the VaJJ!

Atum or the Hidden one was in the Plasma Soup aka Nun swimming. Nun is kinda the hidden Mom in the story, but anyway... I will just tell you go read &/or reread the Horus/Set TransAction book.

(Nun)/Atum
Shu/Tefnut
Geb/Nut
Ausar/Auset & Set/Nephthys

NOW I KNOW THIS MAY BE HARD TO SWALLOW BUT...

Psalms 91:4

He shall cover thee with his feathers, and under his wings shalt thou trust: his truth shall be thy shield and buckler.

The idea is wether you call on Yashua or Yehoshua, you are calling to Shu. Shu is the only son of the Heavenly Father in Kemet. It is not Horus and Ausar, Horus would be Shu's Great Grandson.

I know the Egyptologists have given you Ausar, Auset, and Heru as the Trinity, but that is not right. It is Anubis (Anpu), Wusir & Shu. The later cults developed from Osiris himself as a Triune force. Osiris, Horus & Banebdjed -> Osiris' soul, or rather his ba, was occasionally worshipped in its own right, almost as if it were a distinct god, especially in the Delta city of Mendes. This aspect of Osiris was referred to as Banebdjedet, which is grammatically feminine (also spelt "Banebded" or "Banebdjed"), literally "the ba of the lord of the djed, which roughly means The soul of the lord of the pillar. The djed, a type of pillar, was usually understood as the backbone of Osiris. This system eventually became the Quran, that's too much of a story for now...

Shu is the sound of wind and static, the sound of flowing charge.

Recap

Sh'Mot - I did a quick google and found this from. FirmIsrael.com: The Meaning of "Exodus"; The name of the second book, Exodus, gets a little more complicated. It is a Greek word that means "going out" and points to the content of Israel leaving Egypt. However, as mentioned, this book was not originally written in Greek. So Exodus is not its original name! To Hebrew readers, this is the book of Sh'mot, which means "names", as found in the first verse: "Now these are the names of the sons of Israel who came to Egypt with Jacob; they came each one with his household..." -Exodus 1:1

Again, the name of the book comes from its very first verse. To the ancient Hebrew mind, a name was not just a unique or identifying title of a person, but spoke of someone's character or destiny. Moses means drawn out — His life was used to draw Israel out of Egypt. Elijah means The Lord is my God, and His life was marked by bold and uncompromising stands for the Lord. And likely the most important of them all – Jesus means salvation. Names have valuable meaning. Jesus is not there either though... You see what I am saying, codes inside codes. The name would've been something like Shu...

The biblical Judah (in Hebrew: Yehuda) is the eponymous ancestor of the Tribe of Judah, which is traditionally symbolized by a lion. In Genesis, the patriarch Jacob ("Israel") gave that symbol to this tribe when he refers to his son Judah as a Gur Aryeh' גּוּר אַרְיֵה יְהוּדָה, "Young Lion" (Genesis 49:9) when blessing him. In Jewish naming tradition the Hebrew name and the substitute name are often combined as a pair, as in this case. The Lion of Judah was used as a Jewish symbol for many years, and as Jerusalem was the capital of the Kingdom of Judah, in 1950 it was included in the Emblem of Jerusalem.

Names are very important, so who is the super important, master cheat code: Shu the Lion!

I told Shu already...

I told you, I am a Kemetic Student that identifies as a Messianic Jew, that identifies as a Nazarene.

That simplified, means I am a Christian that has accepted Yashua (the light of the world) as my savior.

Shu literally put Osiris on his Back!!! That is what a Grandpa is supposed to look like!

When performing the Church Ritual communion, what do people eat as the Flesh of Yashua? Bread. What is the origin of that bread?

Exodus 25:30

And thou shalt set upon the table shewbread before me alway.

Shewbread, huh?

I think you have enough at this point, you either have eyes that see and ears that hear or....

The Yod, which was my favorite 'letter' in Ivrit, scholars have now identified as the Shut or feather of Shu. The word has become flesh, via the feather & EuMelanin... The Feather of Shu became your Fascia, pronounced like FaceSha. Coated with with EuMelanin the signal attenuating, data processing mysterious 'Black' stuff.

Djhuty spelled by those early Hebrew Scholars Dalet, Aiyn, Tav, Yod... It is pronounced Duaty or Duty (hence Job)... and means he who has the Knowledge. The Yod at the signifies that the knowledge is stored in the Magnetic Field of the Earth (WaveGuide). The key to the Duat, gateway between the Magnetosphere & the Sun. This is said to be one of the sacred names unpronounceable, unpronounceable as to not reveal its secrets to their oppressors.

Ammun, horns of Amun... David kinged himself with the crown of Amun which held a magnificent jewel! Access to the Waveguide...

Here is another one, you need to know that the Neter are encoded in science...

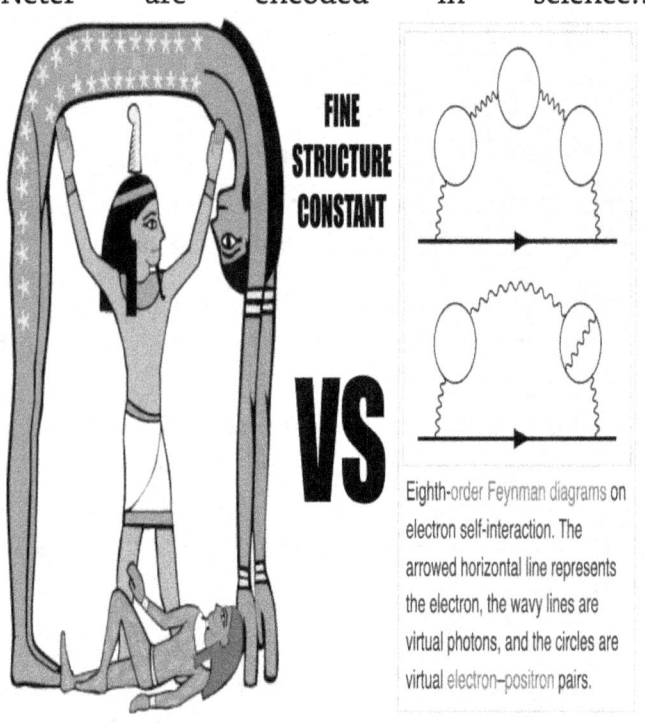

Eighth-order Feynman diagrams on electron self-interaction. The arrowed horizontal line represents the electron, the wavy lines are virtual photons, and the circles are virtual electron–positron pairs.

Geb is the Earth (magnetism)
Shu is the WaveGuide (Electricity)
Tefnut the Magnetosphere (Plasma Lens for Light)

We can either worship correctly or get violated, but once I found the cause and cure for Heart Disease in the Bible... It's UP!!!

Yall know I don't even play with the word cure like that! If you don't know, go get The Observer Effect and the Speak it into Existence books!

Closing...

All of the weird hats and crowns in the Heiroglyphs, are Antennas and Solar Panels! Wake up before the Devil has completely monopolized what God gave you! Please multiply my voice. I beg you. I pray for your salvation and mine.

Lets take back the Air through Prayer and Observance, notice the more we can see, the less the Devil can do. We already 'nosy'... we just need to start praying...

Satan needs synthetic means to commune with God, we have the word turned flesh...

They not like us.

DR.ENQI
Raw Herbal Compounds
PRODUCT GUIDE

Detox Kit:

Kemeluminescence -
NRF2, YEAST, FUNGUS, FAT SUPPORT;
Bladderwrack, Yarrow, Cascara Sagrada, Moss, Happy Tree, Madagascar, Periwinkle, Mayapple, Pacific Yew, Cloves, Amla, Coriander, Black Walnut, Kelp, White Pine Bark, Horny Goat Weed, Milk Thistle, Tribulus, Bitter Melon, Chaste Berry, African Pygeum, Cinnamon, Gynesylvestre, Hemp, Pau D Arco, African Bird Pepper, Cinchona Bark, Chinese Senega Root, Biden Pilosa, Houttuynia, Licorice, Skullcap, Scute Root, Ginseng, Rehmania, Er Bu Shir Tao, Bugleweed

Swadj Momatomix -
Marrow & Electromagnetism Support / Rich in Hydrogen, Phosphorus, Aromatic Amino Acid
Phosphorus, Nettles, Wild Lettuce, Hydrogen, Plant Enzyme, &Alkaloid+ MATRIX

Antiviral Kit -
Antiviral
Antifungal
Antibacterial
mtDNA Protector
The most comprehensive organic antiviral kit ever assembled to fight viral infection and improve recovery

Antivirals -
Exogenous & Endogenous Pathogen Support
Cilantro, Celery, Chaparral, Olive Leaf, Oregano Leaf, Black Walnut, Lysine, Tyrosine, Thyme, Cleavers, Hyssop, Bladderwrack, Ginger

Antiviral Nutrient -
Pathogen Suppression Support
Manganese, Rosemary, Hydrangea, Bilberry, Rhizome Rei

Antiviral Oil -
Immunglobulin & Antibody Support
Oregano, Peppermint, Tea Tree, Cinnamon, Hyssop, Thyme, Clove, Ginger

Calcium -
Muscle & Bone Support
Blood Pressure, Insulin Control, Nerve Function, Muscle Contraction
Kelp, Calcium, Sesame, Cloves

Chromium & Vanadium -
Glucose Tolerance Factor & Eyesight Support
Fat Loss , Insulin Metabolism , Hydration , Muscle Integrity , Energy
Chromium, Fenugreek, Vanadium, Bitter Melon, Gymnema Sylvestre

Copper -
Pigment System Support
Cardiovascular Key, Heart Beat Nutrient, White Blood Cell Reg
Copper, Cilantro, Cloves, Milk Thistle

Iron -
Heme & Magnetism Support
Electron Circulation, Digestive System, Thermogenesis, Brain Power
Iron, Yellow Dock, Stinging Nettles, Chaparral

Magnesium -
Energy & Light Metabolism Support
Muscle Function, Energy, Builds ; Proteins/Enzymes/Hormones , DNA Repair
Blue Vervain, Burdock, Parsley, Magnesium

Muscle Drip -
Children/Adults Multivitamin & Bone Tendon Compound
Blood Oxygen, Breakdown Lactic Acid, Builds Blood Cells Faster, Cleans Lymphatic System
Elderberries, Cherries, Sea Moss, Stinging

Nettles, Horsetail, Lily of the Valley, Bladderwrack, Bromide, Melatonin, Phosphorus, Boron, Calcium, Strontium

Muscle Plants -

Children + Adults Multivitamin & Muscle/Joint Compound

Gout, Autography, Enhanced Healing, Arthritis, Remove Stones

Elderberries, Cherries, Bugleweed, Hombre Grande, Blue Vervain, Chaparral, Ginseng, Rhodiola, Boswellia, Eluethero, Melatonin, Phosphorus, Magnesium

Selenium -

Immune Plasma Support

Thyroid Health, Cancer Suppression, Mental Health, Tumor Suppression

Selenium, Burdock, Bladderwrack, Sarsaparilla

Swadj Momatomix -

Marrow & Electromagnetism Support / Rich in Hydrogen, Phosphorus, Aromatic Amino Acid

Phosphorus, Nettles, Wild Lettuce, Hydrogen, Plant Enzyme, &Alkaloid+ MATRIX

Zinc -

Skin & Enzyme Support

Anabolic Boost, Immune System Nutrient, Stem Cell Health, Gene Support

Rosemary, Chlorella, Sage, Zinc

Watermelanin -

Nootropic, Dopamine, Muscle Recovery, Nourish Your Pineal Gland, DMT Support
Raw Organic Non-GMO Black Watermelon Seeds Lupulin

Anabolic Hormone Help -
Anabolic Hormones, AMPK & Circadian Support
Jiaogulan, Wild Lettuce, Tribulus, Longjack, Maca & Pollen Blend

Histonic -
Histone Sirtuin Support
Grape Skin, Resveratrol, Tyrosine Analogue, Japanese Knotweed

Ocean Steak -
Vegan B12, Carbon, Nucleoside, Protein, Nucleotide, Omega 3 & Eye Support
Phytoplankton, Duckweed, Chlorella, Purple Laver, Chondrus Crispus & C60 Lutein, Zeaxanthin, Ocean Pigment Matrix

Chrondris Crispus -
Structured Water Mucus Membrane Support
Copper, Cilantro, Cloves, Milk Thistle

NON GMO Moringa -
Whole Body Nutrition Support
Raw Organic Non-GMO Moringa

Purple Phaze -
Anti-Aging Longevity Support
FoTi, Pumpkin Seed, NMN, Bhringaraj, Biotin, Silica, Tyrosine, Yucca, White Willow Bark,

French Lilac, NAD

Every item on this list, every compound is not only from God but works on the skin from the inside out, what we need now is topical.

Topical = Tropical

Batana is Great but it's expensive and incomplete.

Researchers identify 135 new melanin genes responsible for pigmentation

Date: August 11, 2023

Source: University of Oklahoma

Summary: The skin, hair and eye color of more than eight billion humans is determined by the light-absorbing pigment known as melanin. New research has identified 135 new genes associated with pigmentation. Vitamin D and Vitamin A... I told you nature doesn't wait to be discovered before getting to work! Melanin vs Diabetes as a Ministry & Movement isn't waiting around to save lives... We have saving lives and creating thought leaders for 20 years!!! The thing is science just discovered 135 genes for pigment and melanin, how the F@3$ have the been acting as if.... This is why we rely on Nature, God & our Ancestors.

We are teaching the world, showing the world...

#HealingLooksLikeThis

Most people don't know what the process of Healing actually Looks Like!!!

People judge health by how your skin looks, literally your complexion. Your Complex of Ions!

Complexion - the general aspect or character of something; the natural color, texture, and appearance of a person's skin, especially of the face.

complexion (n.)
mid-14c., complexioun, "temperament, natural disposition of body or mind," from Old French complexion, complession "**combination of humors**," hence "temperament, character, make-up," from Latin complexionem (nominative complexio) "combination" (in Late Latin, "physical constitution"), from complexus "surrounding, encompassing," past participle of complecti "to encircle, embrace," in transferred use, "to hold fast, master, comprehend," from com "with, together" (see com-) + plectere "to weave, braid, twine, entwine," from PIE *plek-to-, suffixed form of root *plek- "to plait."
The Middle English sense is from the old medicine notion of bodily constitution or general nature resulting from blending of the

four primary qualities (hot, cold, dry, moist) or humors (blood, phlegm, choler, black choler). The specific meaning "**color or hue of the skin of the face**" developed by mid-15c. In medieval physiology, the color of the face was believed to be caused by the balance of humors in the body and indicate temperament or health. The word rarely is used in the sense of "state of being complex."

also from mid-14c.

Humor - the quality of being amusing or comic, especially as expressed in literature or speech; a mood or state of mind. **Each of the four chief fluids of the body (blood, phlegm, yellow bile (choler), and black bile (melancholy)) that were thought to determine a person's physical and mental qualities** by the relative proportions in which they were present.

humor (n.)
mid-14c., "**fluid or juice of an animal or plant**," from Old North French humour "liquid, dampness; (medical) humor" (Old French humor, umor; Modern French humeur), from Latin umor "body fluid" (also humor, by false association with **humus "earth"**); related to umere "**be wet**, moist," and to uvescere "become wet" (see humid).

In old medicine, "any of the four body fluids" (blood, phlegm, choler, and melancholy or black bile).

*The human body had four humors—blood, phlegm, yellow bile, and black bile—which, in turn, were associated with particular organs. Blood came from the heart, phlegm from the brain, yellow bile from the liver, and black bile from the spleen. Galen and Avicenna attributed certain elemental qualities to each humor. Blood was hot and moist, like air; phlegm was cold and moist, like water; yellow bile was hot and dry, like fire; and black bile was cold and dry, like earth. In effect, the human body was a microcosm of the larger world. [Robert S. Gottfried, "The Black Death," 1983]

Their relative proportions were thought to determine physical condition and state of mind. This gave humor an extended sense of "mood, temporary state of mind" (recorded from 1520s); the sense of "amusing quality, funniness, jocular turn of mind" is first recorded 1680s, probably via sense of "whim, caprice" as determined by state of mind (1560s), which also produced the verb sense of "indulge (someone's) fancy or disposition." Modern French has them as doublets: humeur "disposition, mood, whim;" humour "humor." "The pronunciation of the initial h is only of recent date, and is sometimes omitted ..." [OED].

For aid in distinguishing the various devices that

tend to be grouped under "humor," this guide, from Henry W. Fowler ["Modern English Usage," 1926] may be of use:

HUMOR: motive/aim: discovery; province: human nature; method/means: observation; audience: the sympathetic
WIT: motive/aim: throwing light; province: words & ideas; method/means: surprise; audience: the intelligent
SATIRE: motive/aim: amendment; province: morals & manners; method/means: accentuation; audience: the self-satisfied
SARCASM: motive/aim: inflicting pain; province: faults & foibles; method/means: inversion; audience: victim & bystander
INVECTIVE: motive/aim: discredit; province: misconduct; method/means: direct statement; audience: the public
IRONY: motive/aim: exclusiveness; province: statement of facts; method/means: mystification; audience: an inner circle
CYNICISM: motive/aim: self-justification; province: morals; method/means: exposure of nakedness; audience: the respectable
SARDONIC: motive/aim: self-relief; province: adversity; method/means: pessimism; audience: the self
also from mid-14c.

We alway have to remember the Greeks were educated by the Egyptians, ie...

Eu - Good, Perfect, Complete

Melanin (Melanos) - Black

EuMelanin is the new way to reference Osiris.

The new usage of humor came from state of mind, which came from your health, based in the balance of fluids.

#StayWet

The herbs and bitters are to condition the internal fluids, the BleuMagick conditions the body's waters, we are now going to top it off with #SkinFood, to make sure you can #StayWet.

Please read &/or reread Lymphatic Immunity, Mitochondria Water, HydroChemistry...

You've got to learn everything you can from these books about Water. Then you will be ready to apply these principles and practices, Kitchen Chemistry, Orthorexia & this book espouse! Truth be told, these 3 books are plug and play immediately... but the further study is what refines you. You have to disappear sometime and come back stronger. You don't that by consuming new information in your absence.

The biggest difference between those in the Rat Race, and those who aren't, is Priority of Consuming New Information. Reading.

What does it mean to you that, the other pigments in the skin control Melanin production?

What does it mean to you that, they just discovered 3000 new types of Neurons?

Neurons are either specialized Melanocytes or Melanocytes are specialized Neurons, they have discovered 3,000 new types Naga.... Wake Up!!! Either your whole body is a Brain or a Heart! The Heart has all of these cells, Neurons, Nerves, Melanocytes etc...

What does it mean to you that, they just discovered 135 genes that are associated with Pigment? Is that 135 genes for the Skin? Brain? Heart???

My goal is to make #SkinFood inexpensive enough for you and your family to use twice a day, that way we can #StayWet.

In L'Goat we explain in detail how water builds what it needs, with the right conditions. First thing is well, Water. Second thing obviously is Retinoids, to fertilize the soil. Then of course Sunlight...

You are the Dust of the Ground, Divine Soil, remember that the ground exerts pressure on seeds. You need the proper exercise, to create that mechanical pressure to stimulate growth and regeneration. Mechanical Pressure helps to circulate Magnetism, this is the #BleuMagick of #ElectroMagneticTissue.

If this is your first book of ours that you have read... God Bless Eu, or maybe this isn't your first book but you haven't read PiezoElectroChemistry, please do that...

You are the Fruit of Melanin, your S.elf O.rganizing U.niversal L.ight.

Soul is the Fruit of PhotoVoltaic Pigment.

The Bible has many allusions to the **Sun** and its 12 houses, but they replace the Sun with the Son. The Bible is full of Truth, full of Science, you just have to know what your looking at. Kemetic Science is a mixture of Photochemistry, Photobiology & Acoustics. The key is remembering that Light and Sound only exist in your mind. The electromagnetic spectrum you are used to seeing is deceptive, so I have taken the liberty to provide you with a more straight forward version in this book.

ElectronVolts - an electronvolt (symbol eV, also written electron-volt and electron volt) is the measure of an amount of **kinetic energy** gained by a single electron accelerating from rest through an electric potential difference of one volt in vacuum, 1 eV equal to the exact value $1.602176634 \times 10^{-19}$ J, a unit of energy or work, **the work** required to move an electron through a potential difference of one volt. 1 eV would correspond to an infrared photon of wavelength 1240 nm or frequency 241.8 THz.

4-12 ELECTRON VOLTS = UVC 100-320 NM DEATH WM

3.9 ELECTRON VOLTS = UVB SKIN 290-320 NM DISEASE JW

3.4 ELECTRON VOLTS = UVA 320-400 NM EYE DISEASE SW

.001 ELECTRON VOLTS = FAR INFRARED 1000000 NM BEYOND THIS POINT IS RADIO WAVES

.4 ELECTRON VOLTS = NEAR FAR INFRARED 3000 NM

.8 ELECTRON VOLTS = MEDIUM INFRARED 1500 NM

1.5 ELECTRON VOLTS = NEAR INFRARED 780 NM

1.7 ELECTRON VOLTS = VISIBLE RED 620-780 NM

2 ELECTRON VOLTS = VISIBLE ORANGE 585-620 NM

2.1 ELECTRON VOLTS = VISIBLE YELLOW 570-585 NM

2.3 ELECTRON VOLTS = VISIBLE GREEN 490-570 NM

2.6 ELECTRON VOLTS = VISIBLE BLUE 440-490 NM

2.9 ELECTRON VOLTS = VISIBLE INDIGO 420-440 NM

3 ELECTRON VOLTS = VISIBLE VIOLET 400-420 NM

We will look at the E.M. Spectrum in terms of Electron Volts, this makes more sense because all of chemistry is based on the movement of electrons. The Sun is the great visible agent of the first cause. This of course means you have to rethink the whole entire ElectroChemistry aka Dr. Sebi vs Dr. EnQi book...

Sound Range: 20 to 20,000 Hz

Voice Range 90 to 255 Hz

Radio Range: 1 hertz up to 3,000 billion hertz. Below Radio is Cellular, Extremely low frequency (ELF) electric and magnetic fields (EMF) occupy the lower part of the electromagnetic spectrum in the frequency range 0-100 kHz. ELF EMF result from electrically charged particles.

Infrared Range 1 Trillion Hertz to ... this is where our body heat is...

TECHNOLOGY EQUIPMENT RECAP

HydroGenes - 1! Proton and 1! Electron.

Melanocytes - Photovoltaic Cells

Neurons/Nerves - Electromagnetic Cells

Fascia - Plasma Medium (misonomered Ether)

Brain - CPU, Inductor

Brainstem - Two-way Adapter for CPU into the Motherboard

Pineal Gland - Receiver, Crystal Tuner & Actuator Arm/Head responsible for Phosphorescence, Thermoluminescence, Piezoelectricity, Birefringence & Harmonic Generation (very much like the otoconia in the ears)

Operating System - Deductive Logic or PQ

Heart - Hydraulic Ram, Turbine (from the Greek τύρβη, tyrbē, or Latin turbo, meaning vortex) and Hard Drive.

Melanosomes - Alternators

Mitochondria - Motors

Myelin Sheath - Insulation

Cytoskeleton - Filaments

Phospholipids - Capacitors, Dielectric Material (lipids in general)

Spine - Piezoelectric, Motherboard, Radio Wave Antenna

RBC - Floppy Discs

Lymph Nodes - Filters, Nodes

Tastebuds - Electronic Scanners

Protein - Transformer

Transformers 'Roll Out' - Conformational Change (Macromolecule Shape Shifting)

Antioxidants - Semi-conductors (especially the selenium based...)

Body Cells - Plasma based Crystal disc, fitted with integrated circuits as well as gates and channels (see Human Cell Membrane and/or Computer Chip)

Nerves & Vessels - 'Copper' wires (CoAxial Cables) and Fiber Optics

Pigment, Nerve & Blood Clusters - Input Devices like a Mouse, Keyboard, etc...

DNA - Piezoelectric, Antenna, Data Storing Inductors.

DNA sub entry **Tissues** - Short Living Stories.

DNA sub entry **Genome** - Substrate & Product, a digital Library (HardDrive) of all your Ancestors have ever seen, said, touched, tasted or heard.

DNA sub entry **Chromosome** - Rewritable Unlimited Storage Books (Folders) of the Library, DNA.

DNA sub entry **Histone** - Writing instrument, encoders and **book spines**.

DNA sub entry **non-coding RNA** - Self Organizing Books Shelves

DNA sub entry **Gene** - Chapters (Files) in the Books, source codes.

DNA sub entry **Messenger RNA** - Protein Information, a Sentence.

DNA sub entry **Codon** - Word (Binary Code there are 2 bonds between each 3 nucleotides representing their arrangement), Amino Acid.

DNA sub entry **Nucleotide** - Letter

DNA sub entry **Nucleoside** - Bits of Information

Collagen Based Tissue - Piezoelectric Inductors

Stomach - Chemical Mixer

Lumen - the SI unit of luminous flux = to the amount of light emitted per second..... or hollow structures in vessels and cells... hmmm????

Eyes - Camera Lens/Charge Coupled Device (CCD), Digital to Analogue Converter, Complex Photovoltaic Cells/Photodetector...

Amino Acids - Fuses that can be almost anything!

Nucleic Acid - Actual Intelligence (self powering too).

N-Type Semiconductors - Selenium or Silica doped with Phosphorus (Alkaline-ish)

P-Type Semiconductors - Selenium or Silica doped with Boron (Acid-ish)

PN Junction - Crystal Lattice Structure Material allowing the flowing of electrons in one direction.

Bone and Fascia seem to be a massive N-Type, P-Type, PN Junction Super computer on it's own... especially if we add in the Piezoelectricity &

Vitamin D!

Rectifier - N-Type + P-Type + PN Junction in Bone

Melanin - CPU Core, Solar Repeater

Human Cell Membrane and/or Computer Chip - A flat semiconducting (crystal) disc or wafer, with integrated circuits (resistors/conductors) and/or gates & channels. We now have to add the filaments into this Crystal Disc we call a Body Cell or Somatic Cell.

Transistors - a semiconductor device with three connections, capable of amplification in addition to rectification.

The location that a virus goes viral in, is called a **Hotspot**? WTH!

Virus - an infective agent that typically consists of a nucleic acid molecule in a protein coat, is too small to be seen by light microscopy, and is able to multiply only within the living cells of a host.

Wait you see that, it is happening again! Host...

See look there is another definition of **Virus** - a piece of code that is capable of copying itself and typically has a detrimental effect, such as corrupting the system or destroying data.

Wait a damn minute! DNA is a piece of **code**... A viral strand of DNA or RNA that can jump host is fully capable in that context of copying itself, one would even argue, that is it's only 'motion'. The detrimental effects of corrupting the system (physical illness) or destroying data (mental illness), can clearly be seen anthropomorphically.

Alien - Virus

Culture - the arts and other manifestations of human intellectual achievement regarded collectively, the customs, arts, social institutions, and achievements of a particular nation, people, or other social group. <u>The cultivation of bacteria, tissue cells, etc. in an artificial medium containing nutrients, a preparation of cells obtained from a culture</u>. The cultivation of **plants**.

Going Live - become operational.

Live Stream - a live transmission of an event over the internet, transmit or receive live video and audio coverage of (an event) over the internet.

The Web - Arachnoid Mater

Download - copy (data) from one computer system to another, typically over the internet, an act or process of downloading data.

Upload - transfer (data) from one computer to another, typically to one that is larger or remote from the user or functioning as a server, an act or process of downloading data.

Data - facts and statistics collected together for reference or analysis; the quantities, characters, or symbols on which operations are performed by a computer, being stored and transmitted in the form of electrical signals and recorded on magnetic, optical, or mechanical recording media.

Host - an animal or plant on or in which a parasite or commensal organism lives. VS

Host - store (a website or other data) on a server or other computer so that it can be accessed over the internet.

Transmission is the act of transferring something from one spot to another, like a radio or TV broadcast, or a disease going from one person to another.

I am highlighting the unknown and proposing we may have some answers! <u>Infection - an infectious disease.</u>

plural noun: infections "a chest infection"

Vs

Infection - the presence of a virus in, or its introduction into, a computer system. What is a computer system?

Computer System - a computer system is a programmable electronic device that can accept input; store data; and retrieve, process and output information.

Pandemic language = Virology/Biology language. The question is, why? The next question is what does that have to do with Dr. Sebi or Robert Becker? The obvious....

Computer System - a computer system is a programmable electronic device that can accept input; store data; and retrieve, process and output information.

Computer System - a single information processor but usually a group of processors that have specified and general computations; grouped by hardware ie... liver cells, lung cells, brain cells etc.. What you think?

Exercise - activity requiring physical effort, carried out to sustain or improve health and fitness.
"exercise improves your heart and lung power"

Exercise - computer training or computer based training.

Exigenetics - Term created by Dr. EnQi for Melanin vs Diabetes research, denoting the control that exercise has over gene expression.

Hydration - the process of inducing gelling, ionizing, dissolution & turbulent flow with activation of cytochrome c (via infrared light).

Resonance - the quality in a sound of being deep, full, and reverberating. "the resonance of his voice"

- The ability to evoke or suggest images, memories, and emotions."the concepts lose their emotional resonance"

- The reinforcement or prolongation of sound by reflection from a surface or by the synchronous vibration of a neighboring object.

- The condition in which an electric circuit or device produces the largest possible response to an applied oscillating signal, especially when its inductive and its capacitative reactances are balanced.

- The condition in which an object or system is subjected to an oscillating force having a frequency close to its own natural frequency.

- The occurrence of a simple ratio between the periods of revolution of two bodies about a single primary.

- The state attributed to certain molecules of having a structure that cannot adequately be represented by a single structural formula but is
a composite of two or more structures of higher energy.

- · A short-lived subatomic particle that is an excited state of a more stable particle.

Induction - the action or process of inducting someone to a position or organization."the league's induction into the Baseball Hall of Fame"

Induction - a formal introduction to a new job or position.plural noun: inductions
"an induction course"
enlistment into military service.

Induction - The process or action of bringing about or giving rise to something."isolation, starvation, and other forms of stress induction" the process of bringing on childbirth or abortion by artificial means, typically by the use of drugs.

Induction - The inference of a general law from particular instances.

Induction - "the admission that laws of nature cannot be established by induction" the production of facts to prove a general statement.

Induction - a means of proving a theorem by showing that if it is true of any particular case it is true of the next case in a series, and then showing that it is indeed true in one particular case.

Induction - noun: mathematical induction; plural noun: mathematicals inductionthe production of an electric or magnetic state by the proximity (without contact) of an electrified or magnetized body.

Induction - The production of an electric current in a conductor by varying the magnetic field applied to the conductor.

Induction - The stage of the working cycle of an internal combustion engine in which the fuel mixture is drawn into the cylinders.

Is there anyone reading this that would disagree with our body fitting these definitions, the definitions of a computer?

Man this thought experiment just got a lot more interesting didn't it? MIT and the US Military are different types of receipts huh? Is it possible frequency resonance, spreads disease? Human modems? Can Shedding be a broadcast signal?

Wi-Fi is a wireless networking technology that uses radio waves to provide wireless high-speed Internet access. A common misconception is that the term **Wi-Fi** is short for "wireless fidelity," however Wi-Fi is a trademarked phrase that refers to IEEE 802.11x standards.

Viral shedding is a term for when viruses are replicating or reproducing, the virus is being led out of the host cell where it's replicating or reproducing ... Viral shedding is the expulsion and release of virus progeny following successful reproduction during a host cell infection. Once replication has been completed and the host cell is exhausted of all resources in making viral progeny, the viruses may begin to leave the cell by several methods.

Vaccine - a substance used to stimulate immunity to a particular infectious disease or pathogen, typically prepared from
an inactivated or weakened form of the causative agent or from
its constituents or products.

Vaccine - a program designed to detect computer viruses and inactivate them.

"the rate of use of vaccines for computer viruses is not as high as in the US, Japan, and other countries"

Application - a medicinal substance put on the

skin.

Application - a program or piece of software designed and written to fulfill a particular purpose of the user.

In our thought experiment, if a virus is simply the media for harmful information...

Media - an intermediate layer in the wall of a blood vessel or lymphatic vessel.

Media - the main means of mass communication (broadcasting, publishing, and the internet) regarded collectively.

DOPE - an illicit drug (such as heroin or cocaine) used for its intoxicating or euphoric effects especially : MARIJUANA (dopamine altering)

Dope - a preparation (such as an anabolic steroid, diuretic, or tranquilizer) given to a racehorse to help or hinder its performance

To Dope - In semiconductor production, to dope is the intentional introduction of impurities into an intrinsic semiconductor for the purpose of modulating its electrical, optical and structural properties. The doped material is referred to as an extrinsic semiconductor.

Short Circuit - Cardiac Arrest?

Short Circuit - Multiple Sclerosis (due to loss of insulation)

Overheating - Fever?

Overcurrent - Inflammation

With Infection and Virus included we are onto something.

Current - belonging to the present time; happening or being used or done now.

Current - a body of water or air moving in a *definite* direction, especially through a surrounding body of water or air in which there is less movement.

Current - a flow of electricity that results from the ordered directional movement of electrically charged particles.

Current - a quantity representing the rate of flow of electric charge, usually measured in amperes.

Current - the general tendency or course of events or opinion.

Leakage Current - the unintended loss of energy, gain of resistance or results of faulty/worn out insulation.

Plasma - Electric Currents or Electric Current Carrier

Electric Current - Magnetic Field (AtomSphere) Carrier

Alternating Magnetic & Electric Waves - Light

NeuroTransmitters - Record of ElectroMagnetic Waves produced by Neurons (ElectroChemical Message)

Hormones - Large simple versions of NeuroTransmitters (ElectroChemical Message)

Malware - External Negative Mental Programming

Food - Informative Electronic Batteries

0) Movement and sound create energy from water for basic cellular function, via the EnQi Cycle which includes Mitochondria Water. This system slowly increases as all other energy systems fail.

1) Phosphocreatine - anaerobic (no respiration required), phosphocreatine donates it "phospho" to ADP to recycle ATP. This makes 10 ATP per second, its a 1 to 1 ratio (1 phosphocreatine creates 1 ATP) and this is the jump start energy.

2) Anaerobic Glycolysis - anaerobic (no respiration required), Glycogen &/or Glucose to Lactate, 5 ATP per second, 1 to 3 ratio (1 Glycogen creates 3 ATP while 1 Glucose creates 2 ATP) and this is bulk of the energy we focus on, 9 - 120 seconds.

3) NAD/Cytochrome 1 - aerobic (requires

oxygen), Glycogen &/or Glucose to CO_2/H_2O, 2.5 ATP per second, 1 to 38 ratio (1 Glycogen &/or Glucose creates 38 ATP), 2 minutes up to 2 hours.

4) FAD/Cytochrome 2 - aerobic (requires oxygen), FFA &/or Triglycerides to CO_2/H_2O, 1.5 ATP per second, 1 to 360 ratio (1 Glycogen &/or Glucose creates 360 ATP), 2 minutes up to 2 days.

Food rule of thumb - Resynthesis of ATP of Inverse to Yield, the closer the ratio is to 1:1 the fast it can be recycled.

Muscle rule of thumb - Frequently used muscle is slow twitch, Fast twitch is slowly used (at that's the blueprint).

Electric Power - the **rate** at which work is done or energy is transformed into an electrical circuit. Simply put, it is a measure of how much energy is used in a span of time.

Conductor - a person who directs the performance of an orchestra or choir.

Conductor - a material or device that conducts or transmits heat, electricity, or sound, especially when regarded in terms of its capacity to do this.

Lymphatic System - Watermill

Circulatory System - Generator

Integumentary System - Photovoltaic Diaphragm

Immune System - Antivirus, Malware Scanner, Frequency Filter & Rectifier

Nervous System - Power Transmission and Cellular Communications Lines

Fascia System - HydroElectric Grid, Scaffolding

Respiratory System - Windmill

Windmill - a structure that converts wind power or "air" power into rotational energy or vortex energy, to mill grain. In our case grain is Magnetism!

MAGNETS ARE DEFINED BY GRAINS
MAGNETIC GRAINS ARE DEFINED BY APPLIED
STRESS AND CRYSTAL GEOMETRY
SPM SUPERMAGNETIC
SD SINGLE DOMAIN
PSD PSEUDO DOMAIN
MD MULTIDOMAIN

Reproductive System - Quine (self-replicating programs)

Skeletal System - Piezoelectric Crystal Shaped to produced highly specific frequency under stress, Dynamic Oscillators.

Urinary System - Industrial Wastewater, Return

Flow, Surface Runoff, Urban Runoff Agricultural & Animal Husbandry Wastewater

Digestive System - Massive Inductor

Mouth - Industrial Grinder

Endocrine System - Programmer for Human Cell Membrane and/or Crystal Gel Computer Chips

Human Being - Resonator

Vessels - Pipes

Aromatic Ring - Cyclotron (Particle Accelerator)

Glycation - Corrosion

Exegenetics - Holistic Biomechanics; the purposeful science of combining light, water, diet & exercise to effect DNA.

EnQi's 1st Law of Metabolism - The conversion rate of cholesterol should match the activity of Melanin in the skin. These two systems are designed to be and stay coupled. A dark skin person with low sunlight intake and low exercise is going to die from a Metabolic Complication. The only time Animal Flesh is safe to be consumed by a Eumelanin Dominant person is in times of starvation or extremely high activity.

This Law is a Constant and when broken results in disease every time.

EnQi's 2nD Law of Metabolism - The average rate of applied mechanical stress on the bone electrically stimulating bone marrow, determines the rate of bone deterioration and Red Blood Cell production.

EnQi's 3rd Law of Metabolism - The human body metabolizes Transverse Waves and Mechanical Waves into Electricity. Electricity is the main driver of Biochemistry. Exercise is just as potent a driver of Biochemistry as the Sun.

EnQi's 4th Law of Metabolism - Electron movement and bonding is the Nature of Chemistry. PhotoChemistry and PiezoElectroChemistry are the Primary drivers of Biochemistry.

The Ancients discovered this and created Martial Artforms as a way to clean the Bone, Bone Marrow & Brain. Plaque & Sugar are the top drivers of Brain Disease. The things destroying the Heart are secondarily destroying the brain, and they are the breaking of these Universal Laws.

EnQi's 5th Law of Metabolism - Nutrients are actually substrates that must be transformed via biochemistry to be meaningful. This means that providing your body with lots of nutrition without the Water, Light & Exercise don't work alone.

EnQi's 6th Law of Metabolism - The Body maintains the least amount of bone marrow required to handle blood demand. The marrow is very energy demanding, thus attracting and storing fat for energy, eventually becoming fat itself. Fatty bone marrow is called yellow bone marrow. Yellow Bone Marrow can be reconverted to Red Bone Marrow should the body's demands require it, and the body's resources facilitate it. The primary driver is pressure, hormesis training on the Bones. BMR is heavily driven by Bone Marrow, this means Bone Marrow is a driver if insulin and insulin resistance.

EnQi's 7th Law of Metabolism - The system of pigments throughout the body are for metabolism of Light, actual Soulfood. The Adsorption & Absorption of Photons by Water.

Adsorption - increase in the concentration of a dissolved substance at the interface of a condensed and a liquid phase due to the operation of surface forces.

Absorption - a physical or chemical phenomenon or a process in which atoms, molecules or ions enter some bulk phase – liquid or solid material. This is a different process from adsorption, since molecules undergoing absorption are taken up by the volume, not by the surface.

EnQi's 8th Law of Metabolism - in a diabetic

state, sugar is simply invisible to the body. Sugar is not being "sensed" because it's not being converted to energy. In this state of starvation the body turns on every pathway it has to produce sugar from everything you have in your body, fats and proteins included.

This is the reason that it seems like no matter what you eat or 'don't eat', your blood sugar goes up. It's very frustrating. The only way to make it stop is converting that substrate (glucose) into it's final product (energy). The reception of the actual energy, tells the body to stop producing substrate, we good. This must start in the legs and back, the largest muscles but most overlooked. The legs are particularly punished by sitting for extended periods of time, 3-6 hours straight, for a total over 3/4 the time your awake! The leg circulation atrophies and destroys the nerves, nerves are neurons that need a lot of nutrients!

*You must cross reference any and all protocols; food, exercise, medication etc... with the Constitution book & Declaration of Independence!

MELANIN, RADIATION, AND ENERGY TRANSDUCTION IN FUNGI

Arturo Casadevall, Radames J. B. Cordero, Ruth Bryan, Joshua Nosanchuk, Ekaterina Dadachova

ABSTRACT

Melanin pigments are found in many diverse fungal species, where they serve a variety of functions that promote fitness and cell survival. Melanotic fungi inhabit some of the most extreme habitats on earth such as the damaged nuclear reactor at Chernobyl and the highlands of Antarctica, both of which are high-radiation environments. Melanotic fungi migrate toward radioactive sources, which appear to enhance their growth. This phenomenon, combined with the known capacities of melanin to absorb a broad spectrum of electromagnetic radiation and transduce this radiation into other forms of energy, raises the possibility that melanin also functions in harvesting such energy for biological usage. The ability of melanotic fungi to harness electromagnetic radiation for physiological processes has enormous implications for biological energy flows in the biosphere and for exobiology, since it provides new mechanisms for survival in extraterrestrial conditions. Whereas some features of the way melanin-related energy transduction works can be discerned by linking various observations and circumstantial data, the mechanistic details remain to be discovered.

INTRODUCTION

Melanins are dark pigments that are made by diverse fungi (1, 2). Even fungi that produce white colonies, such as *Candida albicans*, have the ability to make melanins (3, 4). Melanins have elicited considerable interest in microbial pathogenesis because they are important virulence factors for many pathogenic microbes, and their presence is associated with reduced susceptibility to antifungal drugs (5, 6). Melanins are multifunctional molecules that give cells structural strength as well as reduced susceptibility to temperature extremes, heavy metals, and molecules produced by the immune system such as oxygen- and nitrogen-derived oxidants and microbicidal proteins (2, 7–10).

Despite the importance of melanins in biology, their structure remains largely unsolved (11). Melanins are composed of covalently polymerized indole- and phenol-type compounds resulting in a material that is insoluble and acid resistant. Melanin structures exhibit local order with global heterogeneity, thus resulting in an amorphous material. Insolubility combined with an amorphous nature means that the structure of melanin cannot be solved with currently available analytical techniques such as X-ray diffraction. A remarkable property of melanins is that they are stable free radicals and manifest a distinctive electron spin resonance signature that is used in their identification (12). Despite the difficulties involved in working with melanin, considerable progress has been made in eliciting its structure by combining results from several techniques including solid-state nuclear magnetic resonance (11, 13, 14).

Melanin production in fungi is a result of three pathways known as the polyketide, 3,4-dihydroxyphenylalanine (L-DOPA), and l-tyrosine degradation synthetic pathways, which produce two chemically different compounds with similar properties (2, 7, 15, 16). Melanin synthesis involves free radical reactions with toxic intermediates. Consequently, melanin synthesis in both animals and fungi occurs in specialized structures known as melanosomes (17). Melanin can be located internally or in and on the cell wall, where it is closely associated with other cellular components such as lipids, carbohydrates, and proteins, although the nature of these associations is poorly understood due to the inherent difficulties in studying melanin structure. For *Cryptococcus neoformans*, one of the fungi in which the process of melanization is best understood, melanin synthesis is catalyzed by a laccase, and the pigment is exclusively 3,4-dihydroxyphenylalanine-melanin. *C. neoformans* melanin is synthesized in vesicles that are exported to the extracellular space and are assembled in the cell membrane into concentric layers, where the pigment is closely associated with polysaccharide and other cellular structures such as chitin and its derivatives (11–14).

Among the remarkable properties of melanin is its capacity for energy transduction (18). Melanin is unique among biological compounds in that its absorption spectrum reveals absorption of all wavelengths in the UV-visible-infrared spectrum. The capacity of melanin to absorb all these wavelengths is presumably a function of its complex molecular structure, which allows it to interact with these frequencies of light. The ability of melanin to absorb electromagnetic radiation extends into the range of X rays and gamma rays, such that it has a shielding capacity that is approximately half that of lead and twice that of carbon (19). Mice given fungal melanin are capable of surviving lethal doses of gamma irradiation, presumably as a result of the pigment protecting the digestive tract and associated lymphatic tissue (20). The ability of melanin to absorb and convert electromagnetic energy is associated with a variety of biological effects that give melanotic fungi tremendous resilience, which translates into their ability to survive in hostile environments.

MELANOTIC FUNGI INHABIT RADIATION-EXTREME ENVIRONMENTS

Melanotic fungi inhabit some of the most extreme environments known, and it is generally believed that the presence of melanin contributes to survival through a variety of mechanisms. For example, Antarctic black rock fungi include a large number of diverse melanotic species that colonize the rocks of one of the most extreme environments on Earth (21). For the purposes of this article we review the literature reporting melanotic fungi in high-radiation environments. The remarkable finding that the damaged reactor at Chernobyl and surrounding soils host a large population of melanotic fungal species is perhaps the best-known example of fungi living in a high-radiation environment (22–24). The Antarctic highlands, where many black rock fungi thrive exposed on the surface and interior of rocks, is also a high-radiation environment (21), especially during the long austral summer and considering the recent historical weakening of the atmospheric protection in the southern hemisphere through ozone depletion. Notably, black fungi recovered from Antarctica can survive in simulated Mars conditions (25). The melanotic fungus *Ulocladium chartarum* was able to grow under space flight conditions, with a rate of growth that exceeded that achieved on Earth (26).

MELANOTIC FUNGI RESPONSES TO RADIATION

Three lines of evidence indicate that melanotic fungi respond to radiation: kinetic attraction toward radioactive sources, faster growth, and metabolic changes. Although these lines of evidence are not completely independent (e.g., hyphal growth toward a radioactive source and enhanced colony growth both reflect cell growth, which in turn is associated with metabolic changes), they are sufficiently distinct to discuss separately.

Radiotropism

The term "radiotropism" refers to the ability of some fungi to grow toward radiation sources (27, 28). This phenomenon was first reported among fungal species colonizing the damaged reactor at Chernobyl and involved the migration toward and degradation of carbon containing "hot particles" (28). Radiotropism was triggered primarily by gamma radiation, but it is possible that both alpha and beta radiation could be "sensed" by the fungi as well. Although the mechanism and purpose of fungal cell migration toward a radioactive source remain unknown, the careful work documenting this phenomenon indicates that fungi respond to radiation and suggest intentional migration and attraction to this energy source.

Enhanced Growth

In 2007 our group reported differential growth effects on melanotic and nonmelanotic strains of *C. neoformans* and *Exophiala* (*Wangiella*) *dermatitidis* exposed to gamma radiation such that the melanin-containing strains grew faster than albino strains (29). This phenomenon was subsequently confirmed and expanded (see below) by another group (30). Enhanced fungal growth in melanotic fungi in response to high-energy electromagnetic radiation has also been reported for other fungal species. Large increases in colony-forming units (>100-fold) were observed for *Alternaria alternata* exposed to gamma radiation (**31**). Exposure of *Aspergillus versicolor* to radioactive nuclides increased hyphal length and spore germination, phenomena that were attributed to enhanced growth by radiation (**24**). Hence, enhanced growth of melanotic fungi in high-radiation environments has been observed for several fungal species by at least four independent groups. Analysis of the energy levels that triggered radiation-related fungal growth revealed that whereas low levels of radiation (150 peak kilovoltage [kVp]) enhanced the growth of both melanized and nonmelanized *C. neoformans*, higher levels (320 kVp) triggered enhanced growth only in the melanized cells (**32**).

Metabolic Changes

Analysis of wild-type and albino mutant *wdpks1 E. dermatitidis* strains 'gene expression after exposure to high-energy radiation revealed differential expression of about 3,000 genes involving multiple pathways, such that genes for cell cycles of amino acid synthesis were downregulated, while stress response genes were upregulated (**30**). Of particular interest was the observation that the wild-type (melanotic) but not the albino *wdpks1* strain manifested ribosomal biogenesis genes' upregulation to radiation exposure, leading the authors to suggest the possibility that melanin-derived energy was being used for protein synthesis (**30**). An experiment with three melanotic fungal species in a simulated Mars environment that included exposure to 200 nm UV radiation revealed some initial changes in protein expression followed by adaptation with resumption of normal expression, leading the investigators to conclude that these organisms could survive in that environment (**33**). An analysis of the response of *Hormoconis resinae* to chronic radiation resulting in cumulative doses of 2 to 3 Gy found that radiation was associated with enhanced synthesis of melanin and antioxidant enzymes (**34**).

INTERACTIONS OF MELANIN WITH HIGH-ENERGY ELECTROMAGNETIC RADIATION

For melanin to capture electromagnetic radiation in a manner that is suitable for conversion to biological energy, it must interact with it. Whereas melanin is known to absorb all UV, visible, and infrared frequencies, the phenomena of radiotropism and enhanced growth upon exposure to gamma rays depend on the ability of the pigment to interact with much higher-energy radiation. Such interactions have now been demonstrated by several methodologies. Irradiation of melanin with gamma rays resulted in changed electronic properties as measured by electron spin resonance spectra (20). A demonstration of the capacity of melanin to serve as an energy transduction molecule for high-energy electromagnetic radiation came from the observation that an electric current was produced by a melanin electrode placed in a gamma-ray beam (35).

IMPLICATIONS OF MELANIN-MEDIATED ENERGY TRANSDUCTION

The phenomenon of radiation-induced growth in melanotic fungi was called radiosynthesis in analogy to photosynthesis, by which plants convert light energy into chemical energy that they can utilize for biological processes (29). Supporting this designation was the observation that irradiated melanin was able to reduce NAD to NADH, thus providing a critical link for the conversion of electromagnetic energy into chemical energy that was immediately biologically useful. However, the analogy of radiosynthesis to photosynthesis is only partial, for the latter is understood to include a series of complex reactions that involve the fixing of carbon to synthesize new biologically useful molecules. For fungi, there are some suggestions that they can utilize CO_2 for synthesizing organic molecules, but this topic has not been thoroughly explored. Studies of black rock fungi from Antarctica using 14C suggested that they could directly utilize CO_2 and incorporate carbon into organic molecules (**36**). Earlier studies of the mold *Zygorrhynchus moelleri* provided evidence for the direct nonphotosynthetic incorporation of CO_2 into pyruvate (**37**). Hence, one can imagine a situation in an extreme environment such as Antarctic mountains where the combination of radiation harvesting by melanin combined with some capacity for utilizing inorganic carbon could produce a process akin to radiosynthesis.

MAJOR QUESTIONS AND FUTURE RESEARCH DIRECTIONS

At this time there is experimental

evidence for the following observations: (i) melanin is an energy-transducing molecule with the capacity to absorb a broad spectrum of electromagnetic radiation; (ii) melanotic fungi exposed to high-energy radiation grow faster than exposed nonmelanotic mutants and unexposed fungi, both melanotic and amelanotic; (iii) melanin can interact with high-energy electromagnetic radiation and transduce it to chemical and electrical energy. In aggregate, these observations suggest that melanotic fungi have the capacity to utilize high-energy electromagnetic radiation to sustain some biological processes. This, combined with the possibility that fungi have some capacity to fix carbon, raises the possibility of radiosynthesis. However, we lack information about the detailed processes by which electromagnetic energy is harvested and converted into biologically useful energy. The ability of irradiated melanin to convert NAD to NADH *in vitro* suggests that it may contribute to energy utilization through simple oxidation-reduction reactions that harvest electrons for biological use. However, there is also a possibility that melanin-related energy utilization involves a sophisticated molecular apparatus such as an antennae complex to shuttle electrons from extracellular cell wall-associated melanin to the cell interior.

The biological advantages of melanin-mediated energy transduction are readily apparent. For organisms in extreme environments such as black rock fungi of Antarctica, the ability to harvest some light energy for biological processes could provide a tremendous survival advantage relative to nonmelanotic organisms. However, the exact contribution of this mechanism relative to the energy available from conventional nutritional processes available to fungi from other sources such as degradation of plant matter is unknown. Establishing the importance of this mechanism in fungi is a much more difficult undertaking than showing the importance of photosynthesis in plants, because the latter are totally dependent on sunlight for their survival, whereas fungi can access nutrients from other sources. In this regard, it is noteworthy that experiments showing that electromagnetic radiation enhances growth have relied on measuring growth increments relative to nonirradiated conditions or albino mutant controls rather than establishing an absolute requirement for growth, since the latter criterion has been experimentally difficult. Consequently, establishing the importance of melanin energy capture is an open question that will probably require innovative experimental design.

IMPLICATIONS FOR BIOLOGY AND EXOBIOLOGY

Given the widespread abundance of melanotic fungi in the biosphere, any amount of conversion of electromagnetic energy into biologically useful energy is likely to have a major impact on estimates of planetary energy flows. For example, photosynthesis is estimated to convert approximately 130 terawatts of sunlight into biologically useful energy, an amount that could increase significantly with melanin-related energy conversion. The ability of melanotic fungi to harvest energy would make them autotrophs and place them alongside plants as important contributors to the conversion of solar and physical electromagnetic energy sources into biologically useful energy. Because the energy of the ionizing radiation is several orders of magnitude higher than the energy of white light, even a very inefficient mechanism of its harvesting will still produce energy transfer. The fact that melanin pigments are easy to synthesize and are found in all biological kingdoms raises the tantalizing possibility that melanin-related energy transduction is ancient, predating photosynthesis, and served as a significant energy-harvesting mechanism for early life on Earth, which emerged in conditions of much higher radiation exposure.

Melanin-related energy transduction has immediate implications for exobiology since it implies that this pigment not only confers the capacity for survival in extreme environments but also provides a means for harvesting electromagnetic energy. The survival of melanotic fungi in simulated Mars-like conditions suggests that the resiliency conferred by melanin already provides some Earth microbes with the capacity to survive on other worlds, thus extending the limits of terrestrial life.

ANALYSIS OF MELANIN PROPERTIES IN RADIO-FREQUENCY RANGE BASED ON DISTRIBUTION OF RELAXATION TIMES

P. A. Abramov, S. S. Zhukov, Z. V. Bedran, B. P. Gorshunov, and Konstantim A. Motovilov(B)

Abstract.

Being a family of biodegradable materials with natural origin, melanins are widely used for development of model bioelectronic devices. However, the mechanism of their electric conductivity is still a matter of discussions.

Current study is devoted to the room temperature impedance measurements of pure and copper-doped synthetic eumelanin at different values of humidity in frequency range 0.1–5·106 Hz. To analyze the obtained impedance spectra, we utilize density relaxation times (DRT) methodology. The performed analysis demonstrates an absence of significant difference in relaxation times in the studied materials. At the lowest frequencies, the doped material has about 30 times lower conductance than pure material. Possible origins of the observed phenomena are discussed in terms of copper ions activity as complexing agent for water molecules and semiquinone groups of melanin.

Introduction

Broad use of electric and electronical devices resulted in the huge increase of corresponding waste. According to the forecast [1], this year world economy is going to produce 52.2 million metric tons of e-waste hazardous to environment. Therefore, the devel- opment of techniques that will allow production of biodegradable "green electronics" [2] is the matter of urgent need for modern society. Melanins are one of the perspec- tive families of biodegradable materials that are able to partially substitute conventional semiconductors used in devices with low energy input [3, 4]. Melanins are pigments that are widely spread in nature in the tissues of cellular organisms [5]. Despite the intensive investigations during more than last 70 years, their structural and conductive properties remain controversial to the scientific community. One of distinctive features of melanin is its hydration-dependent conductance [6, 7].

On the basis of melanin, organic transistors have already been built [8, 9], but their properties are greatly inferior analogs. The purpose of the present study was to investigate mechanisms of conduction of melanin in radio-frequency (RF) range, as well as the evolution of its properties change with humidity.

Properties of melanins in the RF range have been investigated by many scientific groups [10–12], however, the results were analyzed exclusively within the framework of a phenomenological approach, and the physical interpretation of them was either not made or raises a lot of questions. Most studies were devoted to investigations of the dielectric properties of melanin films, whose properties are definitely different from those of bulk melanin, due to the different topology and, probably, different oxidation states as a result of film growth.

Conclusions

We have measured the RF impedance spectra of pellets of pure and Cu melanins in a wide range of humidity values at a fixed temperature. It was shown that conductance of the pure melanin is at least 30 times larger than that of M-Cu melanin, at frequencies <1 kHz.

We found that the presence of Cu leads to a lag in the evolution of the impedance with humidity growth, in contrast with pure melanin. This effect is a subject of further research.

We have shown that the frequency window, within which the spectra are well approx- imated by the finite-length Warburg element, shifts towards high frequencies with the increase of absorbed water content.

The results of DRT analysis allow us to claim that the amplitudes of time peaks (10-2–10-6 s) increase with humidity growth from 0% to 12% and that process is reverse with further increase of humidity from 12% to 84% RH.

This is just part of this article... but go read the whole thing if you need to so no one can ever tell you that Melanin is not active with Radio Waves again!

References

1. http://collections.unu.edu/view/unu:6341

2. Irimia-Vladu, M.: Green electronics: biodegradable and biocompatible materials and devices for sustainable future. Chem. Soc. Rev. 43(2), 588–610 (2013). https://doi.org/10.1039/C3CS60235D

3. Mostert, A.B.: Melanin, the what, the why and the how: an introductory review for materials scientists interested in flexible and versatile polymers. Polymers 13(10) (2021). https://doi.org/10.3390/polym13101670. Art. no. 10

4. d'Ischia, M., et al.: Melanins and melanogenesis: from pigment cells to human health and technological applications. Pigment Cell Melanoma Res. 28(5), 520–544 (2015). https://doi.org/10.1111/pcmr.12393

5. D'Alba, L., Shawkey, M.D.: Melanosomes: biogenesis, properties, and evolution of an ancient organelle. Physiol. Rev. 99(1), 1–19 (2018). https://doi.org/10.1152/

physrev.00059.2017

6. Mostert, A.B., et al.: Role of semiconductivity and ion transport in the electrical conduction of melanin. Proc. Natl. Acad. Sci. U. S. A. 109(23), 8943–8947 (2012). https://doi.org/10.1073/pnas.1119948109

7. Jastrzebska, M.M., Isotalo, H., Paloheimo, J., Stubb, H.: Electrical conductivity of synthetic DOPA-melanin polymer for different hydration states and temperatures. J. Biomater. Sci. Polym. Ed. 7(7), 577–586 (1996). https://doi.org/10.1163/156856295X00490

8. Sheliakina, M., Mostert, A.B., Meredith, P.: An all-solid-state biocompatible ion-to-electron transducer for bioelectronics. Mater. Horiz. 5(2), 256–263 (2018). https://doi.org/10.1039/C7MH00831G

9. Mostert, A.B., et al.: Engineering proton conductivity in melanin using metal doping. J. Mater. Chem. B 8(35), 8050–8060 (2020). https://doi.org/10.1039/D0TB01390K

10. Sheliakina, M., Mostert, A.B., Meredith, P.: Decoupling ionic and electronic currents in Melanin. Adv. Funct. Mater. 28(46), 1805514 (2018). https://doi.org/10.1002/adfm.201 805514

11. Reali,M.,etal.:Electronictransportinthebiopigmentsepia Melanin.ACSAppl.BioMater. 3(8), 5244–5252 (2020). https://doi.org/10.1021/acsabm.0c00373

12. Paulin, J.V., et al.: Melanin thin-films: a perspective on optical and electrical properties. J. Mater. Chem. C 9(26), 8345–8358 (2021). https://doi.org/10.1039/D1TC01440D

13. Wan, T.H., Saccoccio, M., Chen, C., Ciucci, F.: Influence of the discretization methods on the distribution of relaxation times deconvolution: implementing radial basis functions with DRTtools. Electrochim. Acta 184, 483–499 (2015). https://doi.org/10.1016/j.electacta.2015. 09.097

14. Saccoccio, M., Wan, T.H., Chen, C., Ciucci, F.: Optimal regularization in distribution of relaxation times applied to electrochemical impedance spectroscopy:

ridge and lasso regression methods - a theoretical and experimental study. Electrochim. Acta 147, 470–482 (2014). https://doi.org/10.1016/j.electacta.2014.09.058

We have over 60 titles on Amazon.com
$Ministerenqi
Patreon.com/drenqi
AmericanHealer.Website

www.ingramcontent.com/pod-product-compliance
Lightning Source LLC
Chambersburg PA
CBHW071017240526
45469CB00006BD/1958